HOW SAFE IS OUR FOOD SUPPLY?

HOW SAFE IS OUR FOOD SUPPLY?

J.J. McCOY

Franklin Watts
New York/London/Toronto/Sydney
An Impact Book/1990

Material on page 31 is from *An Autobiography* by Harvey H. Wiley. Copyright 1930 Bobbs-Merrill Co.; copyright renewed © 1957 by Mrs. Harvey H. Wiley. Reprinted with permission of Macmillan Publishing Company.

Photographs courtesy of: Monkmeyer Press Photo: pp. 13 (Spencer Grant), 83 (Hugh Rogers); The Bettmann Archive: pp. 17, 20, 27, 128; Brown Brothers: p. 24; Food and Drug Administration: pp. 40, 56, 133; UPI/Bettmann: pp. 46, 76, 146; Rothco Cartoons: p. 48 (Tom Whittemore); Photo Researchers: pp. 70 (Joe Munroe), 103 (Ray Ellis), 126 (USDA/Jack Schneider); Peter Arnold Inc.: pp. 81 (D. Cavagnaro), 112 (Hans Muller); Texas Department of Agriculture: p. 93 (Karen Dickey); Humane Society of the U.S.: p. 101 (Dr. Michael Fox); Center for Disease Control: p. 120.

Library of Congress Cataloging-in-Publication Data

McCoy, J. J. (Joseph J.), 1917–
    How safe is our food supply? / J. J. McCoy.
        p.   cm. — (An Impact book)
    Includes bibliographical references (p.   )
    Summary: Outlines the various procedures by which foods are preserved, grown, and stored, demonstrating how some of these methods may be detrimental to human health.
    ISBN 0-531-10935-6
    1. Food adulteration and inspection—United States—Juvenile literature. 2. Food contamination—United states—Juvenile literature. 3. Food additives—Juvenile literature. [1. Food adulteration and inspection. 2. Food contamination. 3. Food additives.] I. Title.
TX533.M374   1990
363.19′2—dc20   90-35043 CIP AC

CONTENTS

HOW SAFE IS OUR FOOD SUPPLY?

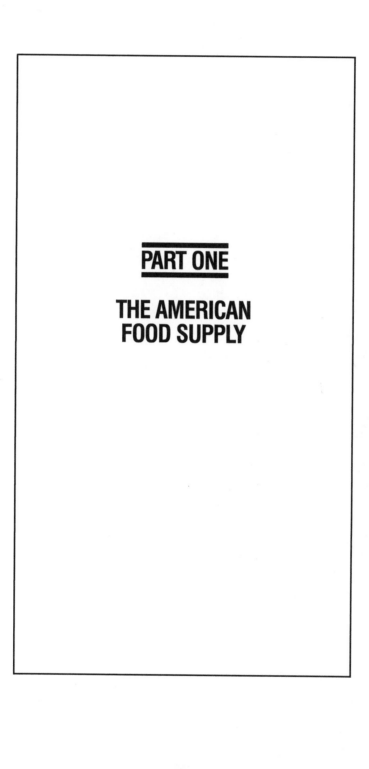

PART ONE

THE AMERICAN FOOD SUPPLY

ONE

THE FOOD INDUSTRY

We North Americans are blessed with a bountiful and varied food supply. Our food chain starts in the sea, the farms and ranches, and the orchards and vineyards. Supermarket shelves, bins, and freezers are crammed with food and food products unknown fifty years ago. Most of us take this plentiful food supply for granted. Few of us think about it or wonder where it comes from. Feeding the nation is complex business.

The American food industry—a multibillion-dollar industry—must provide more than 150 million tons of food each year. Thanks to the efficiency of our farmers and ranchers and our fishing fleets, along with food manufacturers, processors, and distributors, this huge quota is met. In fact, so efficient is our food production system that there is a food surplus. Some of this surplus is stored in granaries and warehouses; some is exported to Europe, Asia, and Africa.

The food industry employs millions of people. Food is collected from all over the country and abroad. On the journey from the farms, ranches, orchards, vineyards, and fisheries, our food travels through a maze of processes, systems, and people.

Packers, processors, inspectors, warehouse handlers, distributors, wholesalers, and retailers all play a role in getting food to our tables.

This gigantic food chain requires a vast storage system. If we did not store food during surpluses, we might face periods of short supply, even famines. In a way, our food storage system is based on a plan used by the biblical Joseph. Joseph, a Hebrew slave of the Egyptians, devised a system whereby grain was stored in good crop years and doled out in bad years. During President Franklin D. Roosevelt's administration, his secretary of agriculture, Henry Wallace, adopted Joseph's plan. Corn, wheat, oats, peanuts, and other food crops were stored in granaries and warehouses to feed Americans in poor crop years. In a sense, our modern food preservation and storage system does what Wallace's and Joseph's did in the past.

In addition to supplying a steady flow of food, the food industry must cater to the tastes of consumers. We have people with different social, cultural, economic, and religious backgrounds. This diversity produces a variety of food preferences. There is another important factor in the modern food chain: convenience.

Supermarkets stock all kinds of convenience and processed foods. There are frozen foods, precooked foods, dehydrated foods, canned foods, pickled foods, smoked foods, and dried foods. Food manufacturers and processors know that their products must be made attractive. Therefore, food is packaged in containers of various shapes to appeal to the consumer. But packaging must satisfy some other important requirements: for example, convenience in handling, safety in storage, and preven-

A wide variety of basic and specialty foods are found in this huge new supermarket.

tion of spoilage. Food manufacturers are constantly developing new food products or improving old ones. And they continue to make more attractive and practical food containers.

The abundance and overall quality of our food supply have helped to raise the standard of living for most Americans. Nutritious foods are available on a year-round basis because of advances in food technology, production, processing, preservation, and storage. Thus, we can say—with some important exceptions to be covered in later chapters— Americans have a wholesome food supply.

But this has not always been true. Before the advent of freezers, modern packaging techniques, and storage facilities, a great deal of food spoiled. To prevent spoilage, eighteenth- and nineteenth-century Americans had to rely on such preservation techniques as smoking, pickling, and canning if they wanted to keep food for a period of time. Even with these preservation methods there was a loss of food and, no doubt, many cases of food poisoning.

Food Adulteration in the Nineteenth Century

Food spoilage was not the only problem consumers had in the nineteenth century. Adulteration of food was a common occurrence. There was no federal pure food law, and those states that did have some food regulations had difficulty enforcing them. Food processors and manufacturers had almost a free hand with their products. And because they did, they frequently adulterated their foods.

The growth and expansion of American cities and towns spawned an organized food industry. This burgeoning industry concerned itself with the growing, processing, handling, storing, transporting, and marketing of food. Some nineteenth-century food processors—like some business operators today—did things the cheapest way to increase their profits. Whenever they could get away with it, they adulterated their products.

There were various ways of adulterating food. Tea is a good example of adulteration practices. Because most tea imported into America came from China and India, there was an excise tax placed on it. This tea tax added to the processing costs. To make up for the tax cost, tea processors stretched their supply by adding dried leaves from ash trees. The ash leaves increased the bulk of the tea supply.

Another trick of the tea processors was to reuse tea leaves. First, tea leaves were collected from hotels and restaurants. Next, the used tea leaves were dried, stiffened with gum arabic or some other thickening agent, and colored to look like new tea leaves. In some cases, lead was used to color tea leaves. Lead, as we know today, is a poisonous element. But no one seemed concerned about possible risks in lead-colored tea in the nineteenth century.

Other foods were adulterated. Farmers and dairies added water to milk to increase the volume. Pickles were given a bright green color by the addition of copper sulfate. Colorful candies were produced by the addition of highly poisonous salts of copper and lead. Many commercially baked breads contained alum, a hazardous substance. And the bright red rinds on some cheeses came from the addition of red lead. And so it went; most nineteenth-century Americans often did not know what was in their food.

In the second half of the nineteenth century, the American food supply, although abundant, was flawed by adulteration, mislabeling, and contamination. The fast-developing food industry included processors who cheated by mislabeling their products. Others risked the health of consumers by adulterating their products. And there was little the consumers could do about it.

The Meat-Packing Industry

Although there were small meat-packing plants scattered around the country, Chicago was the center of the industry. It was an industry that employed hundreds of immigrants in the nineteenth and early twentieth centuries. These slaughterhouse workers were forced to work in unhealthy and sometimes dangerous conditions. Diseased hogs and cattle were slaughtered and made into sausages, hams, cuts of beef, and canned beef. Sanitation was at a minimum or, in some plants, nonexistent. Worst of all was the brutalization of the men who worked on the "killing line," the place where hogs and cattle were butchered. Because of the nature of their work, the killing-line men became callous and unfeeling toward the animals they killed.

There was a loose system of inspections in the Chicago slaughterhouses. It was so lax that diseased and contaminated meat was allowed to be sold. Inspectors simply certified that all diseased meat was kept in Illinois and not shipped to other states. The truth was, however, that spoiled and diseased meat was shipped interstate in violation of federal regulations.

Hogs that had died from swine cholera were

A nineteenth-century woodcut illustrates the working conditions of the men on the killing line in a small meat-packing plant.

slaughtered and sent to a plant in Indiana. Here they were rendered into a fancy grade of lard. Tubercular steers also were slaughtered, despite the prohibition against using such cattle. The meat inspection system was corrupt. Hush money or bribes were paid to the inspectors by the big packinghouses.

"Everything but the Squeal"

Packinghouse workers used to say that all parts of the hog were utilized except the squeal. Nothing went to waste in the Chicago slaughterhouses. If spoiled meat could not be used for canning or for sale as steaks, chops, and roasts, it went into the sausage machines. The awful smell of rotting meat was eased somewhat by rubbing the meat with baking soda. This retouched meat was then sold to charity kitchens serving free meals to the poor of Chicago.

There were other flagrant health violations in the packinghouses. Old sausages, covered with slime or mold, were chopped up and made into "new sausages" after being dusted with borax and glycerine. Meat might slip out of a worker's hands and fall onto the packinghouse floor. Some meat dropped into the dirt and sawdust where workers had spat. Under the watchful eye of a supervisor, contaminated meat was picked up and dumped into the sausage hopper.

The smoking of meat was a time-consuming process. The meat-packers found a way to speed up the process. They preserved the meat with borax and then colored it brown with gelatin so that it resembled smoked meats.

The Packinghouse Jungle

It was the contaminated meat and the brutal working conditions forced on packinghouse employees that prompted Upton Sinclair, a novelist and political activist, to expose what went on in the meat industry. Sinclair's famous novel *The Jungle* told the story of Ukrainian immigrants and the conditions under which they worked in Chicago packinghouses. The novel was published in 1906 and prompted all kinds of reactions.

The meat-packers, angered by the book's portrayal of their industry, launched an attack on Sinclair. They called him an agitator, a muckraker. They ridiculed the book, saying it was pure fiction with no truth in it. The meat-packers did everything they could to reduce the book's influence on the public.

Actually, Sinclair's book is a novel based on fact. The author had done his homework and was able to refute the charges made by the meat-packers. He produced strong evidence supporting the details he had put into his book. He presented fact after fact dealing with the unsafe and unsanitary conditions in the Chicago slaughterhouses. He backed up with hard evidence his descriptions of the brutal and hazardous working conditions. He supplied evidence of diseased and spoiled meats and meat products sold to the American public.

There was one incident in Sinclair's book that he could not substantiate. He described how, now and then, a worker would tumble into one of the huge vats that rendered suet into lard. If a man fell into the lard vat, according to Sinclair, his family was never told the truth. When a man failed to come

THE JUNGLE

UPTON SINCLAIR

The cover from Upton
Sinclair's 1906 novel The Jungle

home, and his family inquired after him, they were told anything but the truth. Usually, the family was told that the man had collected his pay and gone off somewhere, probably worn out by his economic troubles.

The meat-packers pounced on this account. They called it another example of Sinclair's stretching the truth. The workers who had told Sinclair about this kind of accident did not back him up because they feared losing their jobs. But Sinclair insisted it was true. Everything else he had written about the horrible conditions in the meat-packing industry was later proved to be true. When officials of the United States government inspected some of the packing-houses, they found Sinclair's descriptions and statements to be accurate.

Another incident that pointed up the unsafe meat supply occurred during the Spanish-American War in 1898. American soldiers fighting in Cuba were fed spoiled beef that became known as "embalmed beef." It was later said that more soldiers died from spoiled beef than from the bullets fired by the Spanish soldiers.

Sinclair's book and the embalmed beef episode served to arouse Americans. They called for reforms in the food industry and a federal law to protect the public from contaminated and adulterated food. One man who led the fight for pure food laws was Harvey H. Wiley, a physician who became the chief of the Bureau of Chemistry in the United States Department of Agriculture.

TWO

THE BEGINNING OF THE PURE FOOD MOVEMENT

Dr. Wiley and some members of Congress began a campaign for a federal law regulating the manufacture, processing, and distribution of food. The first pure food bill was introduced into Congress in 1889 by Senator A. S. Paddock of Nebraska. The bill passed in the Senate but was rejected by the House of Representatives. Other similar bills met with the same fate. The demands for a pure food law were regarded by members of Congress, the administration, and the food industry as coming from cranks, crackpots, and do-gooders who lacked business sense.

The strongest opposition to a national food law came from the steadily growing food industry. The industrial growth of the country after the Civil War had brought all kinds of changes. Increased and centralized populations—that is, in cities—brought a demand for preserved and canned foods. This demand for more food led to growth in food manufacturing and processing. To meet the demands for food and to protect their profits, food processors chose the most convenient methods of preserving their products—sometimes without full consideration of food safety. Adulteration and mislabeling of food were often the result.

As the demand for regulation of the food industry grew, food manufacturers and processors attempted to unite against regulation of their industry. They resented interference and opposed laws they feared would affect their processing and operating methods.

A common food preservative in use in the early part of the twentieth century was benzoate of soda. Other additives included alum, sulfuric acid, and salicylic acid, all now considered to be hazardous to health. Vegetables were canned with copper sulfate to give them a fresh green appearance. Sulfur dioxide was used in bleaching sugar because sugar processors believed the American public wanted only white sugar, not the natural brown. Sugar and molasses processors were staunch opponents of any attempt to regulate their business. Similarly, the manufacturers of the artificial sweetener, saccharin, resisted efforts to enact a pure food law.

Gradually, the pure food movement—led by Harvey Wiley as chief of the Bureau of Chemistry and by other concerned citizens—gained momentum. *Collier's Weekly,* a widely read publication, supported the campaign for a pure food law. William Allen White, editor of the *Emporia Gazette* in Kansas, banned fraudulent food advertising in his newspaper. However, the *American Food Journal,* a Chicago-based publication, was on the other side of the issue. The *Journal* printed articles that refuted the claims and charges of pure-food advocates. The *Journal* also promoted the use of benzoate of soda, a chemical that Dr. Wiley stated was hazardous to human health.

The *Journal* and other opponents of a pure food law claimed that Dr. Wiley and other pure-food advocates were out to "destroy American business."

Dr. Harvey H. Wiley in his laboratory

One of the strongest and most vocal of the food law opponents was the Western Packers and Canned Goods Association, later to be part of the National Canners Association.

Dr. Wiley's Poison Squad

Shortly after becoming the chief of the Bureau of Chemistry, Dr. Wiley instructed his assistants to investigate food safety. One project that drew the attention of the public was the formation of what became known as Dr. Wiley's "Poison Squad."

The Poison Squad was authorized by a special act of Congress. The purpose of the squad, according to the act, was to enable the secretary of agriculture (Wiley's superior in the department) to investigate the "character of food preservatives, coloring matter, and other substances added to foods; to determine their relation to digestion and health, and to establish the principles which should guide their use."

Dr. Wiley called for young volunteers to serve on his Poison Squad. Out of a group of employees in the Department of Agriculture, Wiley selected twelve for the experiments he had in mind. He wanted young, robust, healthy volunteers with maximum resistance to the harmful effects of food additives. He reasoned that if his Poison Squad showed signs of injury or ill health after being fed food with additives, it followed that children and older people would suffer greater effects.

The members of the Poison Squad were sworn in for a year's service. They pledged to obey all rules and regulations governing the experiment. Wiley's Poison Squad and the experiments with food additives were firsts. Prior to the formation of the Poison Squad, no food additive experiments had ever been performed on human beings.

When the food industry and the press learned about the Poison Squad experiments, they reacted quickly. Wiley was condemned and called a mad scientist for using human beings in such a wild experiment. Government officials, members of Congress, and some people in the Department of Agriculture tried to get Wiley to drop the project. President Theodore Roosevelt—"TR," as he was called—was not in favor of the experiment. In fact, TR did not think a federal pure food law was necessary. He thought the food industry could police itself. All in all, Wiley was pressured by the government to cease the experiments. He was ridiculed and satirized in the press. Strong editorials damned him for poisoning young people in the interest of science.

However, the young volunteers of the Poison Squad did not object to being used as human guinea pigs. They promised to eat and drink only what was served to them in the special dining room set aside for them in the Bureau of Chemistry complex. A physician was on hand to examine them and to provide advice if the experiment proved too harmful to any squad member.

At the start of the experiment, a normal, wholesome diet was prescribed for each member. Members were weighed every day for ten days. If anyone gained weight during that period, the ration was reduced. If someone lost weight, the ration was increased.

After this preliminary period, common food additives in use at the time began to be added to the

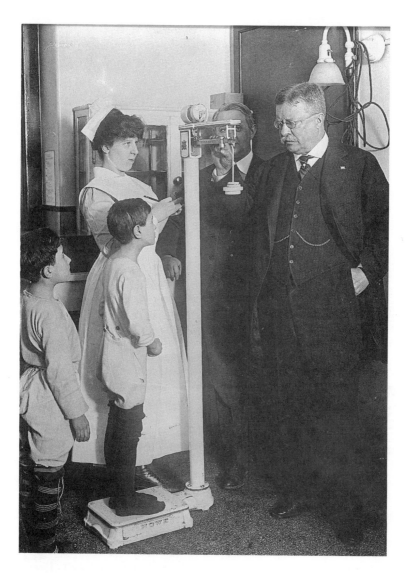

President Theodore Roosevelt weighing schoolchildren in the early 1900s, in a school project aimed at ascertaining the nutritional needs of children.

ration. The first additive was borax. Six members received this additive in the form of boracic acid, and six were fed borate of soda. Since all other variables had been eliminated from the daily diet, the effects of the adulterant on the metabolism of the volunteers could be measured.

When any members of the squad showed signs of digestive trouble, they were given a rest or holiday. Thus, the squad members were able to recuperate for the next experiment. Despite all of the accusations, Dr. Wiley was not a sadist. He allowed no experiment to proceed to the point where it would endanger the lives or the health of any of his dedicated young volunteers.

Wiley's Poison Squad experiments lasted five years. During all of that time, he was censured and vilified. There were demands for his dismissal. His experiments were ridiculed and called a waste of time and money. Above all, he was criticized for trying to regulate the food industry.

The Poison Squad experiments resulted in some definite conclusions about food additives. Preservatives in use at that time—borax, boric acid, salicylates, sulfuric acid, benzoic acid, benzoate of soda, formaldehyde, copper sulfate, and saltpeter—all were in some way injurious to human health.

Dr. Wiley's campaign for a pure-food law and his additive experiments, the furor caused by Upton Sinclair's novel, and the concern of some government officials and a growing segment of the public were all major forces that led to the passage of the Pure Food and Drug Act of 1906.

This law had a stormy passage through Congress. There were strong pressures on Congress by the food industry lobby. Food manufacturers and processors claimed the law was too harsh and that it would ruin business. Scientists employed by the food and chemical industries disagreed with the restrictions on the use of additives. They warned that if preservatives were banned in foods, many Americans would get sick or even die from spoiled and contaminated food.

Some food manufacturers tried to get changes in the law. They wanted a ruling that food manufacturers and processors would have to *know* that an additive was harmful before they could be charged with violating the law. This demand brought up the question of who would determine the nature and extent of any food adulteration. There were no government laboratories equipped to handle such work. Nor did the food manufacturers have any laboratory facilities. Wiley and other pure-food advocates fought against all efforts to alter the Pure Food and Drug Bill. However, attempts to change the bill so that it would be more favorable to the food and chemical industries continued. There were also attempts to kill the bill.

Despite all of the pressures and efforts to water down the bill, it became law on June 30, 1906. Although President Theodore Roosevelt signed the bill, he did so with reluctance. Moreover, he had not supported the bill during its rough passage through Congress. Once the bill became law, most food manufacturers and processors adjusted their operations to conform with the new regulations. Others did not. One group of nonconformers was the sugar processors. They continued to use sulfur dioxide for bleaching sugar.

Dr. Wiley, as chief of the Bureau of Chemistry, and his co-workers were responsible for enforcing the Pure Food and Drug Act. This was no small or easy task. To begin with, there were inadequate funds. There was no large force of inspectors nor machinery for testing all of the food products on the market. Consequently, Wiley had a difficult time enforcing the law. He was obstructed by the food industry and also by members of Congress, the administration, and others, who, for one reason or another, opposed any effort to regulate the food supply. His opponents argued that if the law was enforced according to Wiley's interpretation, the American food industry would be destroyed.

Dr. Wiley and his assistants focused their attention on the major preservatives they knew were indeed harmful to human health. All kinds of arguments and controversies arose over Wiley's attempts to enforce the food law. Catsup manufacturers claimed that they had to use benzoate of soda in their product so that it would not spoil and blow up! Benzoate of soda was one of the additives targeted by Wiley for removal from the food supply. Wiley's scientists proved that catsup could be heated to a point where bacteria and their spores could be killed. Not only would bacteria be killed, but the catsup would not explode on the kitchen shelf.

Saccharin was another problem for Wiley. This artificial sweetener is not a recent discovery. It was discovered in the latter part of the nineteenth century. During Wiley's tenure of office, saccharin was added to canned corn as a sweetener. But the can label did not identify saccharin as the sweetening agent. Wiley told President Theodore Roosevelt that people who ate canned corn thought they were eating sugar-sweetened corn. What they were eating, Wiley said, was an artificial sweetener that, according to his experiments, might be injurious to health.

TR, never one to shrink from a confrontation, disagreed. He told Wiley his warnings were a lot of nonsense. And he told Wiley that his personal physician prescribed daily doses of saccharin.

The pressures to relax the enforcement of the Pure Food and Drug Act became greater. Its opponents got to Wiley's superiors in the Department of Agriculture and managed to have some of his responsibilities curtailed. The increased opposition and reduction in his authority eventually led to Wiley's resignation from the Bureau of Chemistry after a long and dedicated career. But he did not falter in his crusade for better food protection. And his efforts for safe food did lead a number of states to enact food laws along the lines of the federal Pure Food and Drug Act.

A warning issued by Dr. Wiley in 1912 is still true today. He said: "When we permit business in general to regulate the quality and character of our food and drug supplies, we are treading on very dangerous ground. It is always advisable to consult businessmen and take such advice as they give that is not biased . . . because of the intimate knowledge they have of the processes involved. It is never advisable to surrender entirely food and drug control to business interests."

Despite Dr. Wiley's warning, the federal government did yield to the food and chemical industries in a number of instances. Enforcement of the 1906 Pure Food and Drug Act was weak and in some cases nonexistent, and there were violations of the law. But the public's concern was growing; demands for a stronger and more effective food and drug law began to be heard.

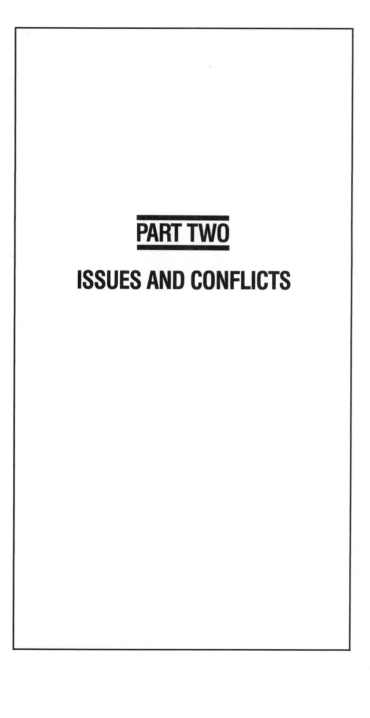

PART TWO

ISSUES AND CONFLICTS

THREE

REGULATING AND PROTECTING THE FOOD SUPPLY

The 1906 Pure Food and Drug Act was soon proven inadequate. Food manufacturers found ways around the regulations. Loopholes were discovered by food manufacturers and processors. It became obvious to food protectionists that stronger regulations were necessary to keep up with the changing times and advances in food technology.

In 1914 the United States Supreme Court ruled that the Pure Food and Drug Act did not require the government to prove that foods containing possibly harmful substances *would* affect the public health. All the government had to prove was that such substances *might* be injurious to the public health. This ruling was important because it allowed the government some leeway in determining the toxicity of a food additive.

Twenty-five years after the passage of the first federal food and drug act, a food and drug agency was established. At first, this agency was known as the Food, Drug and Insecticide Administration. Later the name was changed to the Food and Drug Administration, or FDA, as it is known today. And in 1938, after considerable delays, debates, and changes, a much stronger and more inclusive law, the Food, Drug and Cosmetic Act of 1938 was passed by Congress. It is this law and subsequent amendments that regulate our food supply today.

The new law included the following provisions:

> It authorized standards for identifying foods, their quality, and the amount of food in a container.
> It authorized factory or plant inspections.
> It added the remedy of court injunctions to the already available actions of seizure and prosecution.

Since the enactment of the Food, Drug and Cosmetic Act, there havé been some important amendments. The first standards for food were issued in 1939. These standards were for canned tomatoes, tomato puree, and tomato pastes. Other amendments followed.

A special congressional committee, chaired by Representative James T. Delaney of New York, was assigned the mission of investigating chemical additives in food. Out of the committee's work came three amendments that changed the character and force of the Food, Drug and Cosmetic Act. These amendments were the Pesticide Amendment (1954), the Food Additives Amendment (1958), and the Color Amendment (1960).

All three amendments have had far-reaching effects on the food and chemical industries. Of particular importance to consumers is the Food Additives Amendment. A clause in this amendment states, in part, that "no additive may be deemed safe if it is found to induce cancer when ingested by man or animal."

The amendment also contains a statement to the effect that the ban against cancer-causing additives does not apply to ingredients in animal feeds. That is, the ban does not apply if no residue of such additives is found by approved tests of the food or meat derived from the animals. This provision has been the source of argument and disagreement, especially since the livestock and poultry industries have been feeding animals food containing additives that may be carcinogens.

When these three important amendments were enacted, Americans gained new and potentially powerful safeguards for their food supply. No substance could be added legally to any food unless there had been a prior determination that the substance was safe. Furthermore, food manufacturers and processors were required to conduct research and tests on additives. Enforcement of the first food and drug law had been hampered by a lack of research and testing facilities. That problem was now solved by the new amendments.

The Food, Drug and Cosmetic Act of 1938, with its various amendments, is a complex legal document. It resulted from many weeks and months of discussions, debates, political maneuvering, and legislative compromise. The law is a lengthy list of rules and regulations, but its basic intent is quite clear. In simple terms, the law prohibits the importation into or distribution within the United States of any products that are adulterated or mislabeled.

Adulteration and Mislabeling

Adulteration is the process through which products or materials are made "defective, unsafe, filthy or

[are] produced under unsanitary conditions." The term "mislabeling," or "misbranding," applies to "any statements, designs, or pictures that are false or misleading." The term also includes the failure to provide the required information on a label. Over the years, hundreds of court decisions regarding adulteration and misbranding of food and drugs have been in favor of the consumer.

FDA Center for Food Safety and Applied Nutrition

Although the food and chemical industries are required to test the safety of the additives they use, the FDA does conduct research and testing. The FDA research headquarters, with dozens of laboratories across the country, are located in Washington, D.C. Scientists in the FDA laboratories conduct research and develop standards for the composition, quality, nutritional value, and safety of food and food additives and coloring. Much of the research is designed to improve the detection, prevention, and control of substances that may cause illness or injury when carried by foods or colors.

The FDA Center also coordinates and evaluates surveillance and compliance programs relating to foods and colorings. Food and chemical manufacturers are required to submit what are known as "petitions" to the FDA for the introduction of any new food additive. The Center reviews these petitions. After review, the Center develops regulations or standards for the safe use of the food additives or colors. The Center also collects and interprets data on nutrition and environmental factors that affect the total chemical result presented by any food additives. The Center also maintains a nutrition data bank.

Food additives must be officially approved by the FDA and designated or listed as "generally recognized as safe." This listing is known as GRAS. To determine the safety of additives, a select committee on GRAS was established by the FDA. Scientists from a variety of disciplines—biochemistry, organic chemistry, pathology, physiology, nutrition, food science, oncology, pharmacology, and human and veterinary medicine—were appointed to the committee. Their mission was to evaluate the safety of hundreds of food additives. The committee spent more than a decade investigating the safety of food additives, all under the aegis of the Federation of American Societies for Experimental Biology (FASEB).

The FDA on Food Adulteration

A food or food product is considered to be adulterated if it contains in part or in whole any filthy, putrid, or decomposed material.

If a food or food product has been prepared, packaged, or stored under unsanitary conditions, during which time it becomes contaminated, it is considered to be adulterated.

If a food is a product of a diseased animal, or if an animal has died from a cause other than slaughter, the meat is considered to be adulterated.

If the package or carton or other container holding food is itself composed of harmful or poisonous materials, then the food therein is considered to be adulterated.

*An FDA technician looks at anise
seeds for possible contaminants.*

Any food intentionally subjected to irradiation is regarded as being adulterated (see Chapter 9).

If a valuable ingredient of a food or food product has been omitted in whole or in part, or if any damage or inferiority of a food or food product has been concealed, or if the contents of a box or other container have been packed to mislead by exaggerating the volume or bulk, then the food is considered to be adulterated.

Every new food additive or coloring must be proved safe by scientific research methods accepted by the FDA. The burden of proof is on the manufacturer or processor. The FDA defines how the additive may be used, allowing a safety factor of 100. The amount of additive that may go into a food or food product is limited to 1/100 of the maximum quantity that will produce no toxic effects in animals or human beings. However, the Delaney clause states that chemicals that have been shown to produce cancer in animals and human beings, *even if only at extremely high levels,* may not be used at all in food for human consumption.

Testing Food Additives in Laboratory Animals

At present the most effective way to test the safety of food additives is to give them to laboratory animals. There is no modern equivalent of Harvey Wiley's Poison Squad.

There are four generally used methods for introducing additives into laboratory animals.

By placing the additive in the animal's food or water.

By way of a stomach tube.

Subcutaneously, by injection under the skin.

Intravenously, by injection into a vein.

Additives and
Animal Reproduction

Because it was suspected that some chemical additives not only caused cancer but also played a role in reproduction problems, the FDA, in the 1960s, required manufacturers of an additive to test the product's impact on animal reproduction—that is, for the additive's effects on an animal's production of offspring. This testing is done by mating pairs of mice or rats that have been fed a diet containing the particular additive. The offspring of these pairs are also fed the additive, and when they are mature, they are mated. And so on for x number of generations.

This testing, simplified here, is aimed at learning an additive's effect on animal fertility, production of milk, and the development of embryo and fetus. However, this multigeneration method of testing an additive has drawbacks, according to some scientists.

Some toxic chemicals react on an animal's liver, causing it to produce large amounts of what are called detoxification enzymes. These enzymes then modify the foreign chemical (the additive) to such an extent that the chemical becomes less toxic. The chemical is also excreted more quickly from an animal's body.

What this means is that any animal fed a toxic dose of a food additive for a long period may have a high level of detoxification enzymes. Thus, such an animal could excrete the additive more rapidly than if it were given the chemical for the first time. Such reactions could lead to false readings or findings. One way around this problem is to feed an additive for short periods, say, one to three days during pregnancy. Scientists feel that more food additives have to be tested for their effects on reproduction.

Additives are also subjected to long-term tests. Long-term testing is conducted for the life of the test animal. Animals used in such tests include dogs, cats, monkeys, guinea pigs, rabbits, mice, and rats.

Regardless of the time involved, short-term or long-term, tests may produce results that food industry scientists say cannot be related to human beings. A chemical that may be a carcinogen to a laboratory animal may be nontoxic to human beings and vice versa.

It seems that there is no relationship between systemic toxicity and the capacity to cause cancer. Most scientists agree on this point: there are limitations in testing additives on laboratory animals and applying the results to human beings. This means, then, that food additives may be as safe as we can possibly make them, according to the FDA and food industry scientists.

Mutagenic Food Additives

A further concern for the FDA is the possibility of genetic effects of food additives or chemicals. Some chemicals can cause changes, or mutations, in the genes of animals and human beings. A most dramatic and tragic example of this capability occurred some thirty years ago in Europe. Thousands of deformed children were born to women who had taken the drug thalidomide, a sedative drug similar in structure to Captan, a pesticide used on grapes, while pregnant. The drug was taken to prevent miscarriages.

Human ova and sperm contain DNA, the genetic blueprint, or code, for the development of the offspring. A mutagenic chemical or radiation can mix up the genetic code and produce mutations. For example, a chemical can cause a child to be born with extra toes or fingers, hemophilia, or Down's syndrome. Some mutations may even carry over from one generation to another. Because of the tragic results from the use of this drug, the FDA blocked the use of thalidomide in the United States.

The FDA uses many tests to guard against possibly dangerous additives. One method of testing a food additive for a possible mutagenic effect is to feed laboratory animals large doses. The bone marrow, testes, and spleen are then examined microscopically for signs of damaged chromosomes. The chromosomes are the location of the genetic code. Another testing method is to inject an additive into a cultured cell from an animal. The cell is then examined under a microscope for any changes and signs of genetic damage.

How Effective Are the Tests?

First of all, testing the safety of food additives on laboratory animals is a time-consuming and costly project. Problems exist. One is that testing methods are not as efficient as scientists would like them to be. In addition, both FDA and food industry laboratories test relatively few animals in each specific experiment. The number of animals tested for each additive may be only one hundred. Another question is whether laboratory animals and human beings are equally sensitive to a specific chemical or additive.

Food industry officials claim that experiments employing massive doses of an additive that causes cancer or mutations are invalid. The reason? They say that a huge amount or overdose of *any* chemical may cause cancer or mutations.

Critics of the FDA question the validity of tests required by the FDA. They claim the food industry laboratories—or commercial laboratories engaged by the industry—could tailor the test results to suit the expectations of the manufacturer. Pure-food advocates also say that food industry scientists have a vested interest in producing a favorable report on an additive. In response, food industry officials state that short of outright deceit, required laboratory protocol circumvents this problem.

A Slow-Acting FDA?

For a number of years, the FDA has been criticized for being slow to act when evidence surfaces that a food additive or pesticide may be harmful. One of the severest critics of the FDA and its policies is the consumer group headed by Ralph Nader. In a report entitled *The Chemical Feast*—issued by the Ralph Nader Study Group on Food Protection and the Food and Drug Administration in 1970—a number of charges were made against the FDA. A major charge was that the efforts of the FDA have often been neutralized by powerful special interest groups, such as food manufacturers and trade associations.

Consumer advocate Ralph Nader referring to bottles of fruit juice during a news conference at which he criticized the Food and Drug Administration for trying to kill a rule making fruit juice companies reveal the amount of real juice in their products.

The Nader-group report also criticized the FDA for its failure to regulate the food industry and enforce the law. The report further stated that this failure or reluctance by the FDA to challenge the gigantic and powerful food industry renders our food supply not only suspect but, in many areas, unsafe. The FDA, according to the Nader-group study, defers to the multibillion-dollar food industry. It is claimed that the food industry interests are primary when the FDA decides on new rules and regulations. The Nader-group argues that food industry interests and those of the consumers are not the same; thus, the public should be represented at decision-making sessions.

There are further charges that the FDA has allowed food products to be sold to American and foreign consumers that the agency—and the manufacturers—know to be adulterated and potentially harmful to human health. A recent example of such a situation was the Alar™ controversy in the winter and spring of 1989. Alar—or, to give it its chemical name, daminozide—was sprayed on apple trees as a growth regulator and to increase the fruit's firmness. Animal studies showed Alar to be a carcinogen, yet the FDA did not halt its use until the public furor over the use of this additive.

Another criticism lodged against the FDA is that the agency relies mainly on food industry research and testing. Some critics compare this situation to that of the old warning about putting the fox in the henhouse to guard the chickens.

The Risk-Benefit Argument

According to food industry officials and scientists, the consumer may choose to accept certain risks to

gain the benefits of modern food technology. Risks cannot be avoided when dealing with chemicals. In the case of food additives, a benefit is provided in return for certain risks.

On the other hand, Congress has mandated that no additive shall be placed in any food unless it is proved to be safe. The risk-benefit argument is more or less sanctioned by the FDA. However, it is not fully accepted by some scientists and consumer groups. It may be, as some scientists think, that we may choose to live with additives. But our choice should be an informed one, and it does not mean that those additives that cause cancer and mutations and other ills should not be eliminated from the food supply.

The use of chemicals or additives, pesticide and herbicide residues, arguments of the food industry and of consumer protection groups, and the policies and sometimes inaction of the FDA—all of these are crucial issues and conflicts that affect the health and well-being of all Americans.

FOUR

FOOD ADDITIVES

A hundred years ago, the average American had about eight hundred different foods from which to choose. Today, there are thousands of food products. This great variety of foods is the result of advances in agricultural production and food technology. The food supply is kept available by modern food preservation techniques. These include cooking, drying, salting, bottling, canning, chilling, freezing, dehydrating, smoking, packaging, and— what has become the concern of consumers—the use of chemical additives.

What Is a Food Additive?

Congress has defined a food additive as any substance that, when added to a food, directly or indirectly affects the characteristic of a food or becomes a component of the food.

Canada's Ministry of National Health and Welfare has this definition: A food additive "means any substance, including any source of radiation, the use of which results, or may reasonably be expected to result in it or its by-products becoming a part or affecting the characteristic of a food."

The World Health Organization of the United Nations has this to say about food additives:

A food additive should be safe to use.

A food additive should not be used in any greater quantity than is necessary to achieve the stated effect.

A food additive should never be used with the intention of misleading the consumer as to the nature and quality of a food or food product.

The use of nonnutritive food additives should be kept to a practicable minimum.

Natural Foods versus Food Additives

We hear much about natural or organically grown foods today. Some natural foods contain substances that have no known nutritive value. Others actually contain harmful substances when used in amounts larger than what is considered normal. For example, coffee, tea, and cocoa have no chemical additives but contain caffeine, a substance known to have physiological effects.

Small amounts of arsenic and other toxic metals occur naturally in a number of foods. Hydrogen cyanide, a deadly poison, is found in lima beans, sweet potatoes, peas, cherries, plums, and sugarcane. The amount of cyanide in these foods, however, is too small to cause any serious effects. A person would have to eat enormous amounts of cherries or lima beans or plums to receive a fatal dose. We accept these natural foods because they are not hazardous in the amounts normally consumed.

Intentional Food Additives

When a chemical or other substance is added to a food to aid in processing, for preservation, or to improve the quality of a food, it is known as an *intentional additive.*

Sometimes the purpose of the additive is to improve or maintain the nutritive value of a food. Some foods have only small amounts of vitamins and minerals, because these substances either are not present in the food or are lost during processing.

Additives are also used to maintain the freshness of a food. Our foods last longer on the shelf and in the refrigerator because of certain additives. These chemicals retard spoilage, preserve the natural color and flavor of a food, and prevent fats and oils from becoming rancid or decomposed.

The appeal of a food may also be enhanced by additives. Certain additives make food look better. These are mostly coloring agents; natural and synthetic flavoring enhancers, such as monosodium glutamate (a controversial additive); and artificial sweeteners.

Finally, additives are often used to aid in the processing and preparation of food.

Types of Chemicals Used in Food Processing

There are more than three thousand additives used in food. Each has a specific function.

Preservatives are chemicals added to food to prevent or inhibit the growth of microbes or protect it against decay, discoloration, or spoilage. A number of different types of preservatives are used today, depending on the food product and the organism that might cause spoilage. For example, benzoic acid and sodium benzoate are used to inhibit the growth of molds and bacteria. Sulfur dioxide is widely used in the preservation of dried fruits. Other preservatives are ascorbic acid, calcium propionate, potassium sorbate, and the very controversial sodium nitrate and sodium nitrite.

Antioxidants are used in processing fatty or oily foods, which are susceptible to oxidation or changes in the fat molecules. When oxidation occurs, a result is an off-odor and unwanted taste. Antioxidants prevent this kind of spoilage. Two commonly used antioxidants are butylated hydroxyanisole (BHA) and butylated hydroxytoluene (BHT). They are usually listed on the label of a food or food product.

Acids or acidifiers used in foods, include sodium aluminum phosphate, tartaric acid, monocalcium phosphate, and sodium acid pyrophosphate.

The tart or tangy taste of soft drinks—other than the colas—is produced by the addition of organic acids. These include citric acid from limes, lemons, and oranges, malic acid from apples, and tartaric acid from grapes.

Bleaching agents are another group of additives. When wheat is milled, it has a pale yellow color. However, during storage the yellow fades to white, and the wheat undergoes an aging process. This aging process improves the baking quality of the wheat. The aging process can be speeded up by adding certain chemicals. They include benzoyl peroxide, oxides of nitrogen, chlorine dioxide, and nitrosyl chloride. Some of these chemicals have a bleaching effect; others bleach and age the wheat.

Commercially baked bread is treated with a number of additives. Some of the bread improvers contain small amounts of oxidizing substances, such as potassium bromate and potassium iodate. Only small amounts of these chemicals are used; overuse would result in an inferior bread or other baked product.

Emulsifying agents are used to keep oil-and-water mixtures from separating. They are used in baked goods, cake mixes, ice cream, frozen desserts, and certain candies. Emulsifiers include lecithin, monoglycerides, and diglycerides. They improve the volume, texture, and uniformity of most bakery products. They are also used in candies to maintain uniformity and improve the keeping quality of the candies.

Flavoring agents are important additives as flavor is an essential part of any food. Few people eat what does not taste right. To enhance the flavor of a food product, manufacturers add special substances, such as amyl acetate, benzaldehyde, carvone, ethyl acetate, vanilla, and paprika. All kinds of spices and extractions of spices are used to flavor sausages and prepared meats. Modern chemistry techniques have produced many synthetic flavors that cannot be distinguished from the natural flavor. Almost any natural flavor can be duplicated today, and the results are very convincing. For example, chemists can produce a peach flavor with a chemical known as anisic alcohol. Both kinds of flavoring—natural and synthetic—are widely used in ice cream, soft drinks, baked goods, and candies. Monosodium glutamate, a controversial additive, is sometimes used as a flavor enhancer.

There are more than 1,500 synthetic flavoring

agents available to the food industry. Twenty of them are on the FDA GRAS list. Hundreds more are on what is called the FDA Regulated List. This list contains additives that are allowed in food but are subject to regulation.

No one knows the long-range effects of artificial flavors on human health. Some people can tolerate the artificial flavors; others cannot. For instance, monosodium glutamate, frequently used in Chinese food, has caused various reactions and symptoms in some people, ranging from headaches to difficult breathing. This additive reaction has become known as the Chinese restaurant syndrome. Monosodium glutamate has also caused illness in infants. For this reason baby food manufacturers have stopped using the additive, although it is on the GRAS list.

Food coloring agents create the fresh look and color that most people prefer. Since many foods lack color or fade with time, food manufacturers cater to this preference. Both natural and synthetic food coloring agents are added, especially in processed foods. These chemicals play an important part in making food attractive. However, coloring agents can also be used to hide damaged, inferior, or stale food products. Therefore, FDA regulations specify that food coloring must be used with honesty and discretion.

Red, green, and yellow are the common colors used in food. At one time, Red Dye Number 2 was widely used in candies, soft drinks, hot dogs, pickles, jellies, jams, prepared meats, lipstick, and cough medicine. Research on the dye conducted by Russian scientists in the 1960s suggested that this dye caused birth defects, fetal deaths, and cancer in

*Many of the rich colors we take
for granted in processed foods are
due to the use of color additives.*

laboratory animals. While some questions remain about the reliability of the research, the FDA eventually banned the use of the dye over protests from the food and cosmetic industries.

In the early 1970s *Consumer Reports,* a consumer-oriented magazine, issued a warning about red dyes. The message stated that women of childbearing age, especially those in the first three months of pregnancy, should avoid noncola drinks unless the label clearly stated that the product contained no red dye.

Food Coloring, Flavoring, and Children

Some researchers report that certain additives, particularly flavoring and coloring agents, may cause adverse reactions in children. For example, some artificial flavors and colors are now believed to cause behavioral problems in children. The late Dr. Ben Feingold, former chief of allergy at the Kaiser Permanente Medical Center in San Francisco, was a pioneer in this field of research.

In tests conducted by Dr. Feingold and his associates, some serious effects were seen when susceptible children consumed food or beverages containing artificial flavors or colors. Some children developed hives, or lumps on the skin, accompanied by intense itching. Other children had rashes. Still others had asthmatic reactions with wheezing and coughing. And a number of children showed behavioral changes.

Dr. Feingold's group believes that some artificial flavors and colors may cause learning disabilities. The additives produce a condition known as minimal brain dysfunction (MBD) or specific learning disability (SLD). It was thought that children suffering from this condition were psychotic. However, Dr. Feingold and his associates demonstrated that this was not necessarily so. Some children are simply allergic to artificial flavors and colors.

Children who react to artificial flavors and colors have genetic variations, according to Dr. Feingold's conclusion. They do not, as some observers believe, have abnormalities that predispose them to MBD or SLD.

Dr. Feingold's studies suggest that children with a tendency for overactivity could develop one or more of the following behavioral problems as a result of consuming flavoring and coloring additives:

Poor sleep habits
Fidgeting
Hyperaction
Compulsive aggression
Frustration
Clumsiness
Poor schoolwork

These symptoms formed a pattern of behavior that Dr. Feingold believed to be the result of consuming food or beverages containing artificial flavors and colors. High on the list of foods and beverages that contain such additives are cereals, hot dogs, packaged cakes and cookies, candy, ice cream, fruit punches, ice pops, cheese, and margarine. Popular beverages consumed by children are the quick-mix powdered fruit drinks often called "bug juice." These drinks have both artificial coloring and flavoring.

Supplementary Nutrients

A number of foods have little nutritive value. For many years, people ate cereals without much food value other than as roughage. The processing of cereal grains, such as oats and wheat, removes a large portion of the natural vitamins and minerals. In response, partly to address this deficiency, the FDA approved the addition of vitamins and minerals to cereals and other foods. The agency formulated standards for the minimum and maximum levels of necessary nutrients, such as thiamine, niacin, riboflavin, iron, and other trace minerals. The regulations also permit the addition of vitamin D and calcium. The major cereal companies voluntarily add vitamins and minerals to their products in amounts to correspond to the amounts in natural cereals before processing.

Vitamin A is added to margarine, and vitamin D is added to both fluid and canned milk. Some manufacturers add vitamin D to Gorgonzola and bleu cheeses as a replacement for the vitamin D lost in the use of low-fat milk and in the cheese-bleaching process. Other nutrients added to food include amino acids, baker's yeast, kelp, and potassium iodide. All are beneficial.

Other Intentional Additives

Tannin, gelatin, and albumin are used as clarifying agents in the production of vinegar and certain beverages. They clear sediment in the liquid and remove tiny particles and minute traces of minerals such as copper and iron.

Sugar substitutes are added to foods for people

who want to reduce the caloric level of the sweet foods they eat. Saccharin, aspartame, and cyclamates are artificial sweeteners used in foods and beverages. Saccharin and the cyclamates have been highly controversial additives. The cyclamates were banned in 1969, but the FDA announced in 1989 that the agency would lift the twenty-year-old ban in a few years. These artificial sweeteners are discussed in Chapter 7.

Sequestering agents are substances that remove or set apart other substances or ingredients. They are used to prevent unwanted effects of metallic ions in food products. The sequestering agents do this by combining inactive complexes with the metals in a food.

Humectants add moisture to food and are used to prevent the drying of certain candies. Shredded coconut, a favorite ingredient in many candies, tends to harden with time. The addition of a humectant, such as glycerine, propylene glycol, or sorbitol, keeps the coconut soft and pliable. Certain waxes and gum benzoin are used to coat candies to give them a moist glaze or luster. Boxed candy is usually treated in this way.

Other additives include various chemicals added to fruit and vegetables to improve their texture and appearance. As most cooks know, canned tomatoes, potatoes, and sliced apples tend to become soft or fall apart. To prevent this from happening, small amounts of calcium salts are added to fruit and vegetables to make them firm.

Sodium nitrate and sodium nitrite—two chemicals believed to cause cancer—are widely used in the curing of meats. Meat processors say that these chemicals produce and stabilize the pink or red color most people want in meat. And meat-packers defend their use of these two chemicals, arguing that sodium nitrite and sodium nitrate also kill bacteria.

True—but their consumption can be harmful to infants under six months of age, according to infant mortality reports. These infants are very susceptible to a condition with the long name of methemoglobinemia. It is caused by the blood's inability to carry oxygen. What happens is this: the blood hemoglobin is converted into methoglobin because of the action of sodium nitrate or sodium nitrite. The discovery of this reaction led to the banning of the two chemicals in baby foods. However, they are present in a number of foods, such as hot dogs, bacon, olive loaf, and other prepared meats. Food scientists consider them among the most toxic chemicals in our food supply.

Researchers have found that sodium nitrite has the ability to form nitroamines—compounds known to be carcinogens—teratogens, and poisons. A carcinogen is a substance that causes cancer; a teratogen is a substance that causes birth defects. Significant levels of nitroamines have been found in cured pork, bacon, some fish, and sausage.

Under present FDA regulations, cured meat and some other foods may contain up to 500 parts per million (ppm) of sodium nitrite, which is considered to be a negligible risk. The exception is smoke-cured fish, in which only 10 ppm of sodium nitrite is allowed to enhance the color. However, the Canadian Ministry of Health prohibits the use of sodium nitrite in any fish or fish products.

Food Additives and Labels

Labels on food products are important to consumers. The facts, or data, on labels describe the product and are supposed to tell what ingredients are in the food. Labels should also give the nutritive value of a food, the manufacturer or processor, and, in most cases, when the product should be used. Labels may also provide information vital to the health of consumers. For instance, a label might list ingredients or substances in the product that some consumers must avoid, such as the percentage of fat, cholesterol, and salt.

Not all labels list everything in the product. The amount of data on a label varies according to FDA regulations. However, all food labels must contain at least the following information:

Name of the product.
Net content or net weight, including the liquid in a can, jar, or bottle.
Name and place of business of the manufacturer, packer, processor, or distributor.

In general, all ingredients must be listed on the label. They must also be identified by their common name. The ingredient present in the food in the largest amount, by weight, must be listed first on the label. Other ingredients follow in a descending order according to their weight.

All additives in a food or food product must be listed. If colors or flavors are used, FDA regulations allow the use of such terms as "artificial color" or "artificial flavor" or "natural flavor." But these terms do not tell the consumer what color or flavor agent has been added to the food. There is one exception: the use of the artificial color known as Yellow 5 must be stated on the label. This coloring agent can cause allergic reactions in sensitive persons.

Standardized Foods

Some food products are exempt from the ingredient-listing regulations. These are known as standardized foods. They are products to which the FDA has applied "standards of identity." Foods in this category must contain certain ingredients if they are to be called by a certain name. For example, catsup is a standardized food product. So is mayonnaise. All bottles of catsup and jars of mayonnaise, no matter who makes them, are supposed to have the same ingredients, although formulas may differ. There are about three hundred standardized food products.

New guidelines governing the clarity of labels were proposed in 1990 by the United States Department of Health and Human Services. If these measures are adopted, practically every food product label—and there are about 20,000 such labels in the average supermarket—will have to be changed.

The Chemical Dilemma

Most people are not chemists and do not know what the chemicals and other additives listed on a label are or why they are in a particular food or product. Therefore, they must rely on the research and advice of the FDA, as well as the reputation of the manufacturer or processor. But how reliable are these sources? In recent years, consumer concern about the safety of the food supply has greatly increased. Consumer protection groups, such as Americans for Safe Food, charge that additives approved by the FDA may be harmful to human health. The FDA and food manufacturers counter this charge with the argument that chemical additives are safe when used in recommended amounts.

Some food industry scientists point out that the human body is made up of chemicals. Thus, chemicals are not foreign to the human body. These scientists point out that all of us are consuming chemicals in small amounts; therefore, the human physiological system apparently can absorb them, especially since there have been few, if any, deaths from consuming additives.

But food protectionists say that at the present time nobody can state with certainty that any part per million or even part per billion of an additive is totally safe. On the other hand, it seems that nobody can say that minute levels of an additive will have an adverse effect on an individual or the entire human population.

There are no easy answers. Food additives are an integral part of the American food chain. We have a food supply system that relies on chemicals to provide and maintain a steady flow of wholesome food. And there are scientists who say that it is not possible to eliminate chemical additives without severely disrupting our food production, processing, and distribution.

FIVE

PESTICIDES

The widespread use of pesticides, with residues showing up in food, has caused Americans to take a second look at how we produce our farm commodities. The abundance of our food supply, according to some agricultural scientists, has been made possible by the application of fertilizers and pesticides. This may be true, but there are mounting fears that reliance on pesticides and other farm chemicals may threaten the health of human beings, as well as livestock and wildlife.

What Is a Pesticide?

A general definition of a pesticide is any chemical or compound or mixture used to repel, kill, or destroy a plant or animal pest. There are more than 270 different pesticides in use on American and Canadian farms and ranches.

Pesticide Classification

Pesticides are divided into the following groups:

Avicides are used to control birds that consume or destroy crops and fruits.

Bactericides destroy bacteria.

Fungicides are used to control fungi on fruit and vegetables.

Herbicides are used to destroy weeds and other unwanted plants.

Insecticides are used to kill insects and other arthropods, such as ticks and spiders.

Miticides repel or kill mites, small parasitic arachnids.

Molluscicides repel or kill snails and slugs.

Nematicides are used to destroy nematodes, or threadlike worms.

Rodenticides are used to kill rodents, such as mice and rats.

Toxicity of Pesticides

All pesticides carry the risk of being toxic to animals and human beings. This possible toxicity and the threat it poses to human health has caused consumers to question the widespread use of pesticides.

Environmental groups are concerned about the following pesticides used by American and Canadian farmers:

Aldicarb, also known as Temik™, is a chemical compound known to weaken the immune system of mice exposed to low levels in their drinking water.

Alar is a pesticide applied to fruit as a growth regulator and to maintain the firmness of the fruit. There is some evidence of carcinogenic effect. This pesticide cannot be removed from fruit by washing or peeling; it is translocated into the flesh of the fruit. Although banned for use in the United States, it is being sold abroad.

Benomyl is a soil fumigant or disinfectant used before planting. It has been shown to cause cancer and genetic mutations in laboratory animals.

Captan is a fungicide used on fruit and vegetables. It is known to cause cancer in laboratory animals.

Parathion is an insecticide that may cause birth defects in chicks and ducklings. There is evidence that it also causes cancer and nervous system damage.

Do We Need to Use These Pesticides?

This question is uppermost in the minds of many American and Canadian consumers. Some agricultural scientists say that American agriculture—the best in the world—could not have reached its high production peaks without the use of pesticides. Actually, the use of poisonous substances to control plant and animal pests is not a recent agricultural technology. Colonial farmers—Washington and Jefferson among them—applied poisons to their cotton and other crops to control insects. Farmers in the nineteenth and early twentieth centuries resorted to poisons to control insects and other pests. Copper sulfate, arsenic, lead, and even powdered tobacco leaves were used to fight plant pests. Paris green, a poisonous emerald green powder, was a common insect-killer used by American farmers in the first half of the twentieth century.

The potent pesticide DDT came on the scene during World War II. It was used by the armed forces to control disease-bearing insects and arachnids, such as mosquitoes, lice, and ticks. Soon its use as an agricultural pesticide became widespread, and farmers beleaguered by boll weevils, corn borers, cabbage worms, and other pests used it with abandon. DDT became a major agricultural tool throughout the world.

This synthetic pesticide was the forerunner of a host of pesticides and agricultural chemicals that appeared after World War II. Chemical manufacturers sang the praises of the farm chemicals. Agricultural research centers and agricultural extension services, by way of pamphlets and their county agents, advised farmers on the use of DDT and other farm chemicals.

Then came the rude awakening. DDT had played havoc with the environment. The environmental destruction and loss of wildlife caused by the wide and indiscriminate use of DDT and other pesticides was eloquently and graphically described in the late Rachel Carson's book *Silent Spring,* published in 1963. This book, like Upton Sinclair's *The Jungle,* served as a warning and awakened the American public to the possible danger to their environment from the overuse of pesticides.

Regulation of Pesticides

The environmental hazards posed by DDT and other farm chemicals—and the public concern over their use—led to the enactment of the federal Insecticide, Fungicide and Rodenticide Act in 1972, when DDT was banned. But this chemical compound is so hard to break down that residues are still found in soils and food in those parts of the country where farmers relied heavily on this pesticide. And birds that consumed DDT in their food chain, such as the bald eagle and peregrine falcon, are only now recovering from the decimation the pesticide wrought on their populations.

The Insecticide, Fungicide and Rodenticide Act has been amended several times. The law now requires that all pesticides be registered with the Environmental Protection Agency (EPA) and with any states that have a similar law. Pesticide manufacturers must present to the EPA enough data to describe potential toxic effects before the product can be sold. The manufacturer must also provide instructions for the proper use of its pesticide. The EPA is supposed to screen all pesticides to determine if the claims made by the manufacturer are valid. The Food and Drug Administration (FDA) also has a responsibility as far as the enforcement of regulations of pesticides is concerned. However, critics of these two federal agencies charge that the EPA and FDA have been lax in protecting the public from the hazards of pesticides.

Use of Pesticides

Each pesticide must be used in strict compliance with the instructions on the label. Any deviation or omission from the label instructions constitutes a misuse of the pesticide. Violators are subject to civil or criminal penalties. The pesticide label, according to federal and state laws, is a legal document that can be used in court.

Under federal law, all pesticides must be classified as either restricted- or general-use products. Restricted-use pesticides may be bought and used only by certified personnel or applicators, or by per-

Agricultural pesticide spraying on crops near Phoenix, Arizona

sons working directly under the supervision of a certified applicator. General-use pesticides may be used by any farmer, gardener, horticulturist, or other person, without restrictions, unless otherwise printed on the label.

Not all of the pesticides in use have been adequately tested as possible causes of cancer, genetic mutations, and birth defects, according to critics of the EPA and FDA. Many of the pesticides commonly applied on American farms were registered with the federal government before the establishment in 1972 of the EPA. Actually, many active pesticide ingredients sold today were registered before premarket testing regulations were developed. Thus, these older pesticides have not met the tightened registration requirements applied to newer chemicals.

Pesticide Levels in Food

Amendments to the Food, Drug and Cosmetic Act require that maximum allowable residue levels of pesticides be established for each chemical compound used on a specific crop. The levels set by law vary for different crops.

Canada has similar regulations regarding the use of pesticides. As in the United States, all pesticides must be registered. The registration agency is Agriculture Canada. Pesticide applications are reviewed by the Health Protection Branch of the Canadian Ministry of Health, and residue levels are set. Manufacturers must provide certain information: plant and animal metabolism studies, amount of the pesticide to be applied and the frequency of application, toxicity studies, and methods of determining pesticide residues in food.

Testing for Toxicity

Current methods for testing the toxicity of pesticides include the following procedures:

1. Feeding the pesticide to laboratory animals.
2. Exposing an animal's skin to the chemical and then measuring the absorption rate through the skin and into the bloodstream.
3. Allowing the laboratory animals to breathe in the vapor from a pesticide.

(The use of animals for this and other research is a very emotional and controversial issue that is outside the scope of this book. In addition, there is some question as to whether the results obtained from laboratory animal testing is valid when applied to human physiology.)

Toxicity Levels
of Pesticides

The toxicity level of a pesticide is usually stated as LD 50, which means lethal dose 50, and LC 50, or lethal concentration 50. These terms represent the amount or concentration of the toxicants or active ingredients that will kill 50 percent of a tested population of laboratory animals. The toxicity value of a specific pesticide, based on a single dose, is stated in milligrams of pesticide per kilogram of animal body weight, or mg/kg. The value may also be stated in parts per million (ppm).

The LD 50 and LC 50 levels are important in comparing the toxicity of different ingredients in pesticides. The lower the LD 50 or LC 50 reading, the greater the ingredient's toxicity. A high reading indicates a low toxicity level. Pesticides must be used in accordance with the directions on the label.

Pesticides with a high toxicity level must have DANGER and POISON printed in red letters on the label, along with a skull and crossed bones prominently displayed. In 1984 a new regulation ordered the inclusion of the Spanish term for danger, PELIGRO.

Moderately toxic pesticides must display the word WARNING and the Spanish equivalent, AVISO, on the label. Slightly toxic pesticides must carry the word CAUTION (and the Spanish CAUCION) on the label. Regardless of the wording on a pesticide label, it must be remembered that all pesticides are hazardous to human beings and animals.

Recent amendments to the federal Insecticide, Fungicide and Rodenticide Act are aimed at removing older pesticides from the market. This action followed a report that approximately 80 percent of the pesticides in use had not been adequately tested for their ability to cause cancer. The report also stated that 90 percent of the older pesticides lacked sufficient information as to their ability to cause genetic mutations. Also, 65 percent of the earlier pesticides offered no data on their ability to cause birth defects, and 45 percent provided no data as to possible reproductive effects, such as sterility.

Alar and Children

The highly publicized Alar controversy in 1989 provided impetus to the public concern about food safety. It also provided ammunition for pure-food advocates and consumer protection groups, who charged that the EPA and FDA were failing to protect the public from the dangers of pesticides.

Alar had been used on about 15 percent of the nation's apple crop. The EPA and FDA had evidence since the early 1970s that Alar was a potential carcinogen. The FDA wanted to ban the pesticide but was prevented from doing so. The reason? The agency's scientific advisory panel ruled that there was not enough scientific evidence for such an action. Later it was learned that seven of the eight panel members were serving as consultants to the chemical industry at the time of the panel's ruling on Alar.

The Natural Resource Defense Council (NRDC), a private organization dedicated to maintaining a quality environment, issued a report on Alar. This report stated that Alar was a hazardous pesticide that sometimes broke down into unsymmetrical dimethylhydrazine (UDMH), which is under review as a possible carcinogen. Children were at risk because they ate more apples than did adults. According to the NRDC report, the risk of cancer for children was more than four hundred times greater than is acceptable under federal law.

When the public learned about the dangers of Alar, there was a demand for its removal. Many schools stopped serving apples, apple sauce, and apple juice in their cafeterias. This action brought loud protests from apple growers, especially those in the state of Washington, a major apple-producing center. Television and press coverage of the controversy added fuel to the Alar fire.

Sixty Minutes, a popular news-oriented television program, featured a segment on Alar and apples, called "A Is for Apple." Dr. Jack Moore, a deputy administrator for the EPA, was interviewed on the program. He was asked why, when the EPA had evidence that Alar might be carcinogenic, it was still being used.

His reply, in effect, was that once a manufacturer obtains a license to make and sell a pesticide, it is not easy to remove it from the market. The removal process is long and drawn out. The EPA must hold hearings. And a pesticide manufacturer often sues the EPA. Furthermore, the burden of proof that the pesticide causes cancer lies with the EPA.

Twenty Apples a Day

The Alar controversy went on for weeks. Apple growers charged that the *Sixty Minutes* segment on Alar was helping to ruin their business. Some apple growers and distributors claimed that a child would have to eat twenty apples a day before there was any risk of cancer from Alar residues. Uniroyal, the manufacturer of Alar, defended the safety of its product.

Yielding to public pressure (by this time, consumers were avoiding apples), the government made plans to remove Alar from the market. But Uniroyal voluntarily withdrew Alar. A spokesperson for the company stated that Uniroyal would continue to manufacture Alar and that it would be sold overseas.

Some members of Congress expressed concern when they learned that Uniroyal planned to sell Alar abroad. Because the United States imports fruit and vegetables from the foreign countries that would use Alar, it was quite likely that the pesticide would get back into the American food chain. And many

Senator Steven Symms bites into an apple grown without the use of the pesticide Alar during a Senate Labor Committee hearing in 1989.

Americans questioned the ethics and morality of selling to a foreign country a product regarded as unsafe for home use.

Shortly after the removal of Alar, sixty-five American scientists inserted a full-page ad in the *New York Times*. The banner headline read OUR FOOD SUPPLY IS SAFE! The ad pointed out that 1989 was a year in which an epidemic of fear and hysteria gripped the country and caused Americans to doubt the safety of their abundant food supply.

The ad stated, in bold print, that there is no scientific proof that residues of approved and regulated pesticides had ever been the cause of illness or death of children or adults. In their ad the scientists further claimed that there is no scientific basis for the charge that pesticide residues in food caused cancer in human beings. However, the National Resources Defense Council estimated that 1 child in 4,000 would develop cancer from pesticide residues in food, as opposed to the 1 in 1 million risk proposed by the EPA and FDA.

EPA and FDA Policies and Positions

The Environmental Protection Agency and the Food and Drug Administration are two federal agencies responsible for a quality environment and protection of the public health.

The EPA administers federal environmental policies, conducts research, and regulates various environmental areas. For example, the EPA sets standards and maximum allowances for water pollution, hazardous and solid waste disposal, air and noise pollution, pesticides, and radiation. The agency also enforces the standards and regulations regarding these important environmental factors.

The FDA conducts research and develops standards for the composition, quality, and safety of drugs, food, and food products. The agency is also responsible for the proper labeling and packaging of food and drugs. It conducts inspections of food and drug manufacturing plants and issues orders to recall or cease the selling or manufacturing of a drug or food that is hazardous to human health.

However, both the EPA and FDA have been the targets of criticism by pure food advocates. A major criticism is that these agencies are slow to act when a food or food product has been shown to be hazardous. Another federal agency, the General Accounting Office (GAO), a watchdog agency, has issued several reports highly critical of the FDA's performance and procedures regarding the use of a pesticide. The GAO has pointed out that FDA testing for pesticide levels in food has often been inadequate. This agency reports that some tests failed to accurately evaluate toxicity levels of many pesticides.

Another criticism of the FDA's performance in the food protection area involves pesticide levels in imported foods. The United States imports food from a number of foreign countries, for example, grapes from Chile, tomatoes from Italy and Mexico, and shellfish from Mexico, Ecuador, and Canada. All of these countries use pesticides, some of which are banned in the United States. Congress has repeatedly and publicly censored the FDA for its failure to handle this problem effectively.

The EPA has come under fire from pure food advocates because of what they call "a step backward in protecting farm workers and consumers from cancer-causing pesticides in the fields and residues in food."

Originally, the EPA had set what it called a "zero-risk" with respect to allowing pesticide residues in foods. That meant that there was to be no risk from raw and processed foods. In October of 1989, however, the EPA announced that it was changing to a "negligible risk" standard regarding possible cancer-causing pesticide residues in food. What this meant was that insignificant levels of pesticides would be allowed in raw and processed foods.

State Responsibility

State governments are also concerned about the hazards of pesticides. Most states have to deal with environmental pollution caused by pesticides and other farm chemicals leaching into the soil and groundwater. Many states are also concerned about pesticide levels in the food supply. Thirty states have enacted laws dealing with the pesticide problem.

Some states are working toward reducing the dependence of their farmers on pesticides and other agricultural chemicals. They have established standards and certification procedures for organically grown foods. A number of states have special financial programs geared to establish and promote alternative farming systems that do not rely on chemicals. There are programs that provide marketing aid to organic farmers.

In the early fall of 1989 the association of state agricultural departments adopted a resolution urging all states to enact laws and regulations for organically grown foods. Such measures would define organic food. Some states have adopted such laws and regulations.

The Trend toward
Alternative Farming Methods

Many farmers have responded to the public demand for safer food. They have reduced their use of pesticides and other farm chemicals; some have stopped using them altogether. This is especially true for those farmers who specialize in the production of fruit and vegetables. California is a major producer of lettuce, broccoli, celery, tomatoes, grapes, and oranges. In the past, California farmers relied heavily on the use of pesticides and other chemicals. Now, because of consumer demands and objections by farm labor unions, the use of farm chemicals has been reduced in that state. Consumer groups believe that farmers in other states should cut back on their use of dangerous chemicals.

The trend toward less dependence on chemicals and the production of organically grown food is a major step toward providing Americans and Canadians with a safe food supply. Organic farming is not a new agricultural concept or system. It has been practiced for many years in India and England and has many advocates and practitioners in the United States. Many American farmers and gardeners follow the recommendations and advice in such magazines as *Organic Farming* and *Organic Gardening,* published by the Rodale Press in Emmaus, Pennsylvania.

In a recent report, the National Research Council described alternative farming systems aimed at reducing reliance on agricultural chemicals. While the main advantage of these alternative systems is a safer food supply, the NRC report concluded that farmers who reduce their use of pesticides and

[80]

Organic gardening using terracing in a young orchard in Santa Rosa, California

other chemicals will also lower their production costs. A reduction in the use of agricultural chemicals lessens the impact on an already overburdened environment that, in some areas, has suffered from the use of pesticides and pollutants. Less use of pesticides also reduces the threat to human health. Equally important, supporters of alternative methods claim that there does not have to be a decrease in crop yields or livestock production when the use of farm chemicals is decreased.

However, organically grown food costs more. A trip to a health food store will verify this statement. Why the increase in price? Well, for one thing, there is the supply-and-demand factor. Organic food is not as plentiful as that produced with the aid of pesticides and other chemicals. Another factor is that organic farming requires more labor.

Nonchemical Control of Plant and Animal Pests

The recent National Research Council report dealt with a wide range of biological pest-control methods, how and where they are being used, and their potential as future replacements for pesticides.

For example, the federal government has approved test-spraying of a genetically engineered virus for use on plant pests. Natural viruses have been used for such purposes since just after World War II. However, they have some disadvantages. One is that viruses take several days to kill insects; thus, a plant could be damaged before the insects die. Another disadvantage is that viruses stay in the environment, often attacking beneficial insects, such as honeybees, lady bugs, and praying mantises. Lady

A health food store display
of organically grown produce

bugs and praying mantises, as most gardeners know, kill insects, many of which are harmful to food plants.

Insect neurochemistry is another field of research aimed at providing alternatives to pesticides. Entomologists have learned that the insect brain and nerve trunk dictate the insect's behavior. Therefore, entomologists are exploiting this knowledge. They believe that by altering some chemical in the insect brain, it may be possible to kill insects without using chemicals that are toxic to human beings and wildlife.

Chemical manufacturers are working to develop chemicals that are less toxic. For example, Du Pont, a major chemical manufacturer, has been experimenting with a chemical group known as sulfonylureas. These chemicals, according to the manufacturer, are effective and safe when used at the rate of 1 to 2 grams per acre. Other pesticides are usually applied at the rate of 1 to 2 pounds (450 to 900 grams) per acre. Du Pont scientists point out that the low application rate and low toxicity are important features of these new chemical compounds. Their adoption would reduce groundwater contamination, a major problem stemming from the use of highly toxic pesticides.

Whether all farmers will make changes in their farming methods is problematical. Major changes take time. Also, farmers are accustomed to making independent decisions and may not like to be told what they can or cannot do on their farms. Nevertheless, increased public awareness about the hazards of pesticides and other chemicals has put the food industry on notice. Consumers want safe food, and farmers, food manufacturers, and processors will have to adapt to that demand, according to consumer groups.

Postharvest Pesticides

Meanwhile, we can still expect to find pesticide residues in our food supply. A recent survey by the Florida Department of Agriculture showed that 60 percent of the vegetables tested in that state contained traces of one or more pesticides. However, it is important to know that not all of the pesticide residues in food are the result of overuse or improper use of chemicals on the farms. Pesticides are also used after harvesting. They are applied while food is in storage and transit.

Benomyl, a fungicide, is used to protect food while in storage and transit. Postharvest applications of highly toxic pesticides, such as benomyl, are the targets of pure food advocates and consumer protection organizations. These postharvest chemicals are applied to melons and potatoes. Not all pesticides can be washed off fruit and vegetables. Some processors coat tomatoes and cucumbers with a wax containing a fungicide to keep them from spoiling. That glossy look of cucumbers in some stores is the result of an application of the wax and pesticide mixture. Peeling off the skins may not fully do away with the pesticide residue. Some may be absorbed into the flesh of the fruit or vegetable.

New Proposals for Pesticide Controls

In the fall of 1989, President George Bush announced proposals that would cut some of the red tape that has slowed the efforts of the EPA and FDA in banning or removing hazardous pesticides from the market. The proposals were wide-ranging and controversial. When enacted into law, they would do the following:

1. Make it easier for the EPA to halt the use of a pesticide that may be hazardous to human health while the agency collects more data.
2. Reduce the time it takes to cancel the use of a hazardous pesticide by eliminating time-consuming hearings.
3. Eliminate an existing law that prohibits the use on processed food of any pesticide that causes cancer. This may seem contrary to the intent of safeguarding the food supply. However, the EPA's interpretation of this law is that a pesticide cannot be used if it causes more than one case of cancer in a million people over a lifetime. The EPA considers this law too restrictive.
4. Preempt states from applying more stringent standards for pesticide residues than those set by the federal government.

Environmental protection organizations, as well as some members of Congress, have criticized these proposals as being insufficient and, in some measures, a step backward. Critics of the proposals say that they would only slightly speed up the pesticide-removal process. Furthermore, the proposals would have little effect on how rapidly the EPA would act to suspend or remove a pesticide.

Consumer groups were especially alarmed over the government's plan to revise the "negligible risk" standard; that is, the one in a million cancer cases over a lifetime. This standard would be replaced by one that would give the EPA more flexibility in deciding what is a safe level for pesticide residues in food.

Under the proposed standards, pesticide tolerance would range from 1 in 1 million to 1 in 100,000. Also, the allowable risk could be higher if the benefits of the pesticide were determined to be high. Thus, the familiar risk-benefit factor arises again. Pure-food advocates say that this new standard gives the EPA too much flexibility. They also oppose the measures that would prohibit the states from enacting laws that are stronger than those of the federal government. Although the states have rarely used the power given to them by the Constitution to establish stricter laws, critics of the new proposals say that the states should have the option of passing stronger laws, particularly if the EPA and FDA fail to enforce pesticide regulations.

Our use of pesticides is still a perplexing problem. The doubts and confusion that surround the use of pesticides are reflected in a question asked by Arif Jamal, an African agronomist and pesticide expert. Mr. Jamal has wondered how a great country like the United States can ban a pesticide for home use, yet let it be manufactured and sold to developing countries. "Are we supposed to be more resistant?" asked Mr. Jamal.

SIX

HORMONES AND ANTIBIOTICS

Americans consume great quantities of meat each year. Steaks, hamburgers, pork chops, beef and pork roasts, and fried, broiled, and roast chicken are popular foods in the American diet. The high protein value of meat has been widely publicized by the meat industry and nutritionists. In recent years, however, the method of producing the country's meat supply has been questioned by scientists, legislators, and consumer protection groups.

Their chief concern is the use of hormones and antibiotics in livestock and poultry production. Hormones are chemical substances released into body fluids by a gland such as the pituitary gland. Antibiotics are substances produced by certain microorganisms, which, when used in dilution, inhibit the growth of or destroy bacteria or other microorganisms. For nearly thirty years, livestock raisers have produced weight gains in steers and heifers (young cows) by implanting natural or synthetic hormones in the animal's body. And antibiotics are added to animal feed to control disease.

Hormones in Beef and Pork Production

The FDA permits the use of hormones as growth stimulants only after they have been approved for food-producing animals. Although their use has been opposed by consumer groups, livestock producers say that the use of hormones not only yields leaner meat—and thus less cholesterol—but also helps to keep meat prices down. Prices are lower because animals gain weight faster and with less feed.

There are two classes of hormones used to increase growth in livestock. One class contains the *endogenous hormones.* These are hormones that are naturally produced in animals and human beings. They include estradiol, testosterone, and progesterone. The second class are the *exogenous hormones,* or synthetically produced hormones. They include zeranol and trenbolene acetate. Not too long ago another hormone was widely used in the beef cattle industry. It was diethylstilbestrol (DES). When evidence surfaced that DES was carcinogenic, it was banned by the FDA.

The United States Department of Agriculture's Food Safety and Inspection Service (FSIS) points out that only a small percentage of any hormones consumed through food stays in the human body. The estimate is 10 percent. The FSIS emphasizes that hormones are found in foods other than meat. For example, soybeans and other legumes contain small amounts of naturally produced steroids.

Minimal Risk

Tests utilizing radioimmunoassay methods have shown that the increase in overall hormone levels in edible portions of hormone-treated cattle is very small. Scientists say that the natural hormones ingested by eating hormone-treated meat are millions of times less than the average person produces in his or her body. However, for those who are concerned about ingesting hormones, this is not a reassuring statement.

To bolster their argument, FSIS scientists point out that a 500-gram (about 1-pound) portion of meat treated with estradiol has fifteen thousand times *less* estradiol than the average man produces in his body—and several million times less than that produced by a pregnant woman. The FSIS further states that even the most susceptible human beings —prepubescent boys, or boys who have not reached sexual maturity—"produce a thousand times more estradiol per day than they would ingest in treated meat."

It is because of the relatively small amount of hormone residue found in meat that livestock producers and the USDA claim there is no serious threat to the public health. When properly used on the farm and in the feedlots—that is, correctly implanted in cattle and not used to excess—hormones are safe, according to the USDA and FDA.

Consumer groups disagree with the government's assessment and claims regarding the use of hormones. They point out that hormones are known carcinogens. Furthermore, farmers and feedlot operators cannot be relied upon to use recommended amounts of hormone or apply proper implantation techniques. This charge was borne out in an investigation by the USDA in 1986. Federal inspectors found that some ranchers and feedlot operators were using more than the recommended amount of hormones. Also, some cattle raisers were implanting hormones in the wrong part of the animal bodies, that is, in muscles and other parts intended for human consumption. As a result of the investigation, a number of ranchers and feedlot operators were cited for violation of the Food, Drug and Cosmetic Act.

The First Alarm

Public attention was focused on the adverse effects of hormone-treated meat by an incident that occurred in Italy in 1981. The *New York Times* reported that the synthetic hormone DES was found in meat in baby food. DES had been banned in the United States in 1979. However, some Italian farmers bought the hormone on the black market and injected extremely high levels, many times greater than appropriate, into the chest muscles of calves.

The hormone soon spread throughout the bodies of the injected calves, but the highest concentration remained in the chest muscles. It was this portion of the calf that was used in the manufacture of baby foods. The effects of the hormones on babies was startling. Some babies who ate the contaminated food underwent physiological changes. For example, some baby boys and girls developed enlarged breasts. A number of infant girls began to menstruate. People were horrified when they learned about these abnormalities. A result of this episode was a European ban on the use of all hormones in livestock production.

European Ban on
U.S. Hormone-Treated Beef

In January 1988, the European Economic Community's Council of Ministers banned the importation of meat from cattle treated with hormones. Although hormones had been banned in beef production in Europe, some livestock producers were using black-market hormones. A scandal arose in West Germany when fourteen thousand calves were injected with what was called a "hormone cocktail." Most of the calves were destroyed.

The EEC ban on hormone-treated beef caused quick reaction in the United States, a major beef-exporting country. American livestock producers and packers faced the loss of more than $100 million in annual beef exports. The ban was called unfair and unwarranted. Cattle producers argued that the danger from hormone residues in meat was minimal and overstated by consumer protection groups. They emphasized that a result of a total ban would be "a black market product that may be very dangerous to human health." Furthermore, the European ban ignored the findings of American and Canadian scientists, which demonstrated that properly used hormones are safe.

Not all beef produced in the United States and Canada is hormone-treated. A number of livestock producers in both countries are raising hormone-free cattle. The beef from these cattle is labeled as "hormone-free beef" and is being marketed in more than one thousand American supermarkets. It is also being exported and sold in Europe with the approval of the European Economic Community Council. The production of hormone-free beef is believed to be an important advance in livestock management. It is hoped that this example of alternative farming will further reduce or eliminate drug and chemical residues in our food supply.

However, many cattle producers still implant their animals with hormones to achieve increased

Hormone-free cattle production in Texas

weight gain and lower costs of production. They, together with the FDA and USDA, maintain that hormones are safe when properly used. And they resent the European ban. So do the meat processors and hormone manufacturers. The annual income from the sale of hormones is estimated at about $50 million. The market is dominated by Syntex, the International Minerals and Chemical Corporation, located in Illinois; and Eli Lily and Company, based in Indianapolis, Indiana.

Policing the Hormone Ban

European authorities admit that the hormone ban cannot be totally enforced. For one thing, scientists cannot detect hormone residues, especially the natural hormones, with enough accuracy to take legal action against violators. For example, estrogen levels in meat are difficult to measure, and the test results are too variable to be reliable. Even residues of the synthetic hormone trenbolene acetate, which are supposedly easier to detect, present problems for the inspectors. They are almost impossible to detect in muscle meat, fat, and kidney tissue.

Canadian officials are also concerned about the European ban on hormone-treated beef. The director of the Canadian Bureau of Veterinary Drugs, Health and Welfare has approved four hormones for use in livestock. Three of them are natural hormones: estrogen, testosterone, and progesterone. The fourth is the synthetic hormone known as zeranol. These four hormones, according to the Canadian veterinary bureau, present no threat to the health of consumers.

But the Canadian government's claims for hormone-treated meat are challenged by consumer groups in that country. Critics state that some Canadian cattle producers might use illegal hormones; therefore, all hormones should be banned. One Canadian cattle producer was caught using a banned hormone product known as clenbuterol.

Hormones and Cow's Milk

Equally controversial is the use of hormones to increase milk production in dairy cattle. The cause of a new furor over the use of hormones in livestock management is bovine somatotropin (BST). It is a growth hormone that is an almost exact duplicate of the natural BST found in a cow's body. At this writing, BST is still in the experimental-use stage, but FDA approval is expected.

In 1980 Cornell University scientists tested the effects of BST on dairy cows. Their tests, and those conducted at other agricultural colleges, revealed that the hormone raised a cow's milk production by as much as 10 to 25 percent. Like the hormones used on beef cattle, BST also improved feed efficiency. This would mean lower costs for both dairy farmer and consumer.

Four American companies plan to market commercial forms of BST. For example, the Upjohn Company will manufacture a waterlike solution of BST for injecting into the hind leg of a cow. Elanco Products Company (a division of Eli Lily and Company) has developed a BST paste that can be injected into a cow's shoulder once a month. Monsanto Agricultural Company and American Cyanamid Company also plan to market a form of BST.

The FDA has approved for human consumption only BST milk from experimental sources. But many dairy cooperatives in the United States have refused to accept BST milk for distribution. Many dairy farmers and farm organizations, such as the National Farmers Union, oppose the use of BST. According to a survey conducted in March 1989 by *Dairy Herd Management* magazine, 75 percent of American dairy farmers were against the use of BST.

In August of the same year, five of the largest American supermarket chains—Safeway, Stop and Shop, Kroger, Vons, and Supermarkets General—announced that they would not handle BST milk. And food processors using milk in their products, such as Borden and Kraft, announced that they would reject BST milk.

BST Effects on Human Health Not Fully Investigated

Opponents of BST charge that the FDA's review of the hormone was inadequate. Samuel Epstein, a professor of medicine at the University of Illinois, warned that public health threats from BST milk have not been thoroughly evaluated. As for the effects of BST on cows, researcher David Kronfeld believes that use of the hormone will place extra stress on dairy cows, animals already hard-pressed to produce more milk.

On the other side, most of the university scientists who have worked with BST say that it is safe. They base their belief on numerous tests and trials. By June 1989 more than 160 BST trials, utilizing 11,337 cows, had been conducted in North America. And 113 trials involving more than 10,000 cows were conducted abroad.

Opposition to the use of BST is mounting. Many organizations are gearing for a battle if the FDA approves BST for commercial use. According to John Stauber, field director for the Foundation on Economic Trends, a Washington, D.C.–based organization, a coalition of dairy farmers and consumers needs to be organized for the impending fight.

Antibiotics in Animal Feed

The addition of antibiotics to animal feed is another cause for concern for American and Canadian consumers. Antibiotics are effective in the treatment of animal and human diseases. They were first mass-produced after World War II. Aureomycin, streptomycin, and chloramphenicol are some of the antibiotics used in animal feed to combat disease. Two major manufacturers of antibiotics for animals are American Cyanamid Corporation and Eli Lily and Company. American Cyanamid manufactures an antibiotic additive known as Aureo SP 250™; it contains several antibiotics.

Antibiotics suppress the growth of bacteria in the animal intestine. What is more, they control diseases in the early stage, when symptoms may not be present. Livestock and poultry producers maintain that antibiotics are indispensable, especially in the modern factory farming system where animals and poultry are kept in crowded conditions.

The advantages of antibiotics in animal feed have been demonstrated in many livestock feeding experiments. There is little doubt that these feed additives keep livestock and poultry healthy and increase their growth rates. However, opponents claim that most of the research on animal feed additives, like that performed by the chemical and pharmaceutical industries, is suspect. The findings and reports may be influenced by the aims of the manufacturers. Proponents counter that the research is verified by government scientists.

Antibiotic residues are found in meat and milk, and may be a health concern. Some people are allergic to antibiotics, especially penicillin. There is also the question of whether people will build up a resistance to the antibiotics they ingest through their food supply, and thus the drugs will become less effective in fighting human diseases.

There is evidence that antibiotics in animal feed create transferable drug resistance in livestock bacteria. And this resistance to antibiotics can show up in human beings as a result of consuming meat from animals fed antibiotics. But pharmaceutical scientists say this is as yet an unproved theory.

In the 1980s more than three hundred physicians, scientists, and biologists signed a petition that warned about the hazards of indiscriminate and wrong use of antibiotics in livestock production. The petition was sent to President Ronald Reagan and his secretary of health and human services, Margaret M. Heckler. Secretary Heckler rejected the petition on the grounds that an "imminent hazard" had not yet been demonstrated.

FDA Position on Antibiotics in Animal Feed

The FDA has approved a number of antibiotics for use in animal feeds. Among them are bacitracin, bambermycin, erythromycin, virginiamycin, penicillin, and the tetracyclines. Penicillin and the tetracyclines were pointed out by opponents as being of major concern to public health. The FDA, in response to the complaints, asked the National Academy of Sciences (NAS) to conduct risk-assessment tests involving low-dose levels of antibiotics in livestock. In February 1989, the NAS reported that it was "unable to find a substantial body of direct evidence that established the existence of a definite human health hazard in the use of subtherapeutic concentrations of penicillin and the tetracyclines in animal feeds." What this wordy sentence seemed to indicate was that it was all right to add small doses of antibiotics to animal feeds.

The government's overall position on livestock feed additives is that they are useful in raising livestock and present little if any hazard to public health. They are an important economic factor in the production of food animals. They represent another major advance in American agriculture, the envy of the world. And as far as any resistance to pathogenic organisms being passed on to human beings is concerned, there is little evidence of this happening, according to government scientists. Any human resistance is caused by overuse of antibiotics in the treatment of human diseases.

Some pharmaceutical scientists believe that it is possible for a person to acquire antibiotic resistance from handling or preparing meat or by direct contact with animal feces. However, this type of resistance transmission is not regarded as a major threat to public health.

Although the prestigious NAS has found no evidence that animal feed additives pose a risk to public health, opponents are not satisfied with the findings and conclusions of the NAS. They argue that bacterial genes are coded for resistance as the result of increased use of antibiotics in animal feeds.

Pure-food advocates point out that animal bacteria may form colonies directly in human beings or may transfer their resistance capacity to other species of bacteria. These antibiotic-resistant bacteria may then colonize in human beings. Also, the continued use of antibiotics in animal feed may ultimately render them useless in fighting human diseases.

Some Troubles with Chickens

Kentucky Fried Chicken, Chicken Nuggets, and other chicken preparations are favorite foods of millions of Americans. The taste of chicken, coupled with its low fat content, make this protein source an important part of the diet of people of all ages.

Chicken, like beef, pork, lamb, and veal, must be produced under sanitary conditions, according to federal and state laws. No adulteration is permitted. The federal Wholesome Poultry Products Act requires inspection of all chicken and chicken parts. If a state does not have an adequate inspection service, then all poultry products within that state must be federally inspected. More than twenty states now conduct their own poultry inspections.

Poultry processors must meet strict sanitation requirements. Inspectors are supposed to keep a constant watch for violations. Any chicken found to be unhealthy or not fit for human consumption has to be removed from the processing line. FSIS and state inspectors check to see that chicken and other poul-

Crowded and unsanitary conditions in chicken farms lead to the growth of harmful bacteria.

try have not been adulterated. Plant laboratories are required to conduct continuous tests on poultry and poultry products to guard against bacterial and chemical contamination.

Despite federal and state poultry inspection programs, unsanitary conditions do exist in plants. And bacteria-laden chickens and chicken parts are sold to consumers. A few years ago several television news programs exposed unsanitary conditions in some poultry processing plants. In one plant, chickens were dropped on dirty floors by careless workers. The carcasses were picked up without being cleaned and sent along the processing line. Further, chickens are an excellent medium for the growth of pathogenic bacteria, especially *Salmonella* organisms. This problem is treated in more detail in Chapter 8.

Antibiotics in Chicken Feed

Antibiotics are added to chicken feed to perform the same function as those placed in livestock feed. That is, they promote growth and help to control disease. However, kosher chickens—those raised and slaughtered under Jewish dietary laws—are not given food containing antibiotics. In fact, no growth stimulant of any kind is used. Also, very strict sanitary measures are followed in rearing kosher chickens.

Perdue Farms Incorporated, whose outspoken president, Frank Perdue, is often seen on television, reports that its chickens are produced under sanitary conditions and with sound animal husbandry practices. Frank Perdue states that "we do not use antibiotics as growth enhancements in any of our

*An inspector for the USDA checking
ducks to make sure that USDA standards
have been met before sale to the public*

programs or in our birds' diets. Instead we rely, only when necessary, on preventive medicines which have been approved solely for animal use by the Food and Drug Administration."

Despite government assurances as to the safety of hormones and antibiotics, the public concern has not been allayed. Many people question the value of and need for these additives. They are not satisfied with the reports and test findings provided by the government and the livestock industry. Finally, a big question remains to be answered: what will be the effects of a lifelong exposure to hormones and antibiotics in meat and poultry products?

SEVEN

ARTIFICIAL SWEETENERS

In the spring of 1989, the Food and Drug Administration announced that it was lifting the twenty-year-old ban on the artficial sweeteners known as cyclamates. In the 1960s, the cyclamates caused one of the biggest health scares of the century. Bladder tumors were found in rats fed this artificial sweetener.

The Cyclamate Controversy

Cyclamate is the product of an accidental discovery. Some fifty years ago, a University of Illinois researcher was experimenting with cyclo-hexysulfamic acid. He was smoking a cigarette at the time and he put the cigarette down where it came into contact with some powdered residue of the chemical. When he resumed smoking, he found that the mouth end of the cigarette tasted sweet. Thus an artificial sweetener was born.

In 1950, Abbott Laboratories filed an application with the FDA for Sucaryl sodium, a cyclamate product. This artificial sweetener, according to Abbott Laboratories, was to be used in foods and beverages intended for diabetic and other persons restricting their consumption of sugar.

Weight control was one of the touted virtues of the cyclamates. But the National Academy of Sciences' Food and Nutrition Board disagreed with this claim. Tests conducted on human beings and rats showed—when the diets were controlled—that foods containing cyclamates had no influence on body weight.

Abbott Laboratories originally intended to market Sucaryl as a drug. But their test methods and reports submitted to the FDA were rejected as too vague and inconclusive. However, the FDA later approved the use of the sweetener because the agency researchers had conducted studies on cyclamates over a two-year period.

Pure-food advocates seized on the inclusive findings of Abbott Laboratories and the FDA study. They thought that more tests were needed to assure the public that this artificial sweetener was safe. The FDA, however, did not order any further tests.

Since the cyclamates had the approval of the FDA, food and beverage manufacturers quickly put them to use. For nearly twenty years, the cyclamates were the major sweeteners in use in the United States and Canada. The National Academy of Sciences had some reservations about the widespread use of the cyclamates. During the twenty-year period in which the cyclamates were the most widely used artificial sweeteners, the NAS issued several warnings about the possible dangers from their use.

The Food and Nutrition Board of the NAS advised that the public welfare should be placed above the "uncontrolled distribution of foodstuffs containing cyclamates." The board pointed out that the cyclamates did more than just sweeten food and beverages. They caused tumors in rats. Besides, nobody knew how the continued use of this artificial sweetener would affect the health of people of all ages. The NAS board recommended that the cyclamates be subjected to further tests under prolonged and varying conditions.

The NAS warning and advice was more or less ignored by the FDA. The agency made no attempt to restrict the use of the cyclamates. In 1962, the NAS board updated its policy on the cyclamates. The board repeated its warning that the public welfare should take precedence over an uncontrolled distribution of food and beverages containing cyclamates.

Cyclamates Placed on the GRAS List

Despite the NAS warnings and growing public concern over the use of cyclamates, the FDA placed the artificial sweetener on the GRAS list. Once on the list, cyclamates could be used in any food product, in any amount, and—as it later turned out—without being listed on the food and beverage labels. Once the FDA stamp of approval had been placed on the cyclamates, manufacturers claimed their products containing the artificial sweetener were safe. After all, they said, the FDA had put cyclamates on the GRAS list of acceptable foods.

When an additive or food is palced on the GRAS list, it remains there until challenged. The NAS warnings were not a direct challenge. They merely indicated some of the shortcomings of this artificial sweetener. Ninety-five percent of cyclamates pass out of the hman body within twenty-four hours, according to test results. Therefore, the manufacturers declared the sweetener to be safe.

In 1966, Japanese researchers reported that cyclamates passing through the body could create a chemical compound known as cyclo-hexylamine (CHA). CHA, acording to the Japanese researchers, was a dangerous chemical compound. The Japanese reports aroused more doubts and concern about the safety of the cyclamates.

Dr. Jacqueline Verrett, an FDA biochemist, discovered a strong relationship between cyclamates injected into chicken eggs and malformation of the embryos in these eggs. In her report, Dr. Verrett stated that calcium cyclamate and its derivative compounds, CHA and dicyclo-hexylamine (DCHA), were specific teratogens, chemical or substances that cause birth defects. No action was taken on Dr. Verrett's report.

Despite new information about the possible adverse effects of the cyclamates, the artificial sweetener remained on the GRAS list. Other disturbing news about the cyclamates added to the confusion and concern. For example, it was believed that regular use of the artificial sweetener could disrupt the effects of coagulants or anti-bleeding drugs. This meant the cyclamates would inhibit vitamin K efficiency. There was also evidence that the cyclamates had some adverse effects on liver function.

The controversy over the use of the cyclamates went on and on. There were many meetings and debates. Manufacturers of the artficial sweetener lobbied for their products. And the public remained confused. Finally, the FDA commissioner issued a statement to the press. It stated that "totally unrestricted use of cyclamates, commony used as artificial sweeteners, is not warranted at this time." The commissioner's statement was not, by any stretch of the imagination, an all-out condemnation of cyclamates. It bypassed the concerns of some researchers and the National Academy of Sciences. What seemed to be implied by the statement was that the public need not worry too much about the safety of cyclamates. But the statement only served to evoke more concern. Food protectionists, the Nader group in particular, charged that politics had triumphed over science.

FDA Restricts the Use of Cyclamates

Later, the FDA announced that cyclamates would be restricted to use in foods only, not in beverages. The artificial sweetener could also be sold as a sugar substitute in tablet or liquid form. All food products containing cyclamates would have to be labeled to show the content of the sweetener in an average serving of the food product. But the labeling requirements met with such resistance from manufacturers and food processors that the FDA never put it into effect.

Doubts about the safety of the cyclamates increased. New findings showed the carcinogenic and teratogenic hazards of this chemical compound. There were demands by consumer groups for the removal of the cyclamates from the GRAS list. Scientists in the United States Department of Health, Education and Welfare called for more and longer studies of the cyclamates. Such studies should determine if cyclamates already consumed by people had any negative effects. There were also recommendations that anyone who had consumed great quantities of food or beverages containing cyclamates should have periodic examinations for signs of developing bladder cancer. Bladder cancer had appeared in laboratory animals fed cyclamates.

Eventually, new findings, plus the demands for a ban on the use of the cyclamates by consumer groups and many members of Congress, forced the FDA to take action. In 1969, the FDA banned the use of cyclamates. Predictably, the food and chemical industries loudly denounced the action. They argued that there was not enough evidence for the ban, that the findings were inconclusive. Also, there was disagreement among scientists as to the hazards of the cyclamates. And the public fears and concerns had been aggravated by scare tactics on the part of the press.

Cyclamates Now Regarded as Safe

After being banned for more than twenty years, cyclamates have been given a reprieve. According to researchers, new evidence collected over twenty years demonstrates that cyclamates, once thought to cause bladder cancer, are harmless The FDA stated in 1989 that it had made a mistake in banning this artificial sweetener. Food and chemical manufacturers said that the FDA yielded to politics and the press. The charge was unfair. It overlooked or ignored the FDA's Congressional mandate: Do not approve any substance known to cause cancer in animals.

But the new findings on cyclamates have not put the controversy or doubts to rest. There are still some questions about the safety of this artificial sweetener. One is the chemical's capacity to cause cell changes. Another is its long-term effect on the human heart and blood pressure.

Saccharin

This artificial sweetener has been in use for more than 100 years. In recent years, however, it has come under attack by some scientists and consumer groups. Saccharin is a derivative of petroleum-based substances. When first synthesized, it was used as a food preservative and antiseptic. Sugar shortages during both world wars prompted consumers of sweets to turn to saccharin.

Saccharin, like the cyclamates, was placed on the GRAS list before the passage of the Delaney amendments to the Food, Drug and Cosmetic Act. Therefore, this sweetener was regarded as safe. The National Academy of Sciences' Committee on Food Protection issued a report on saccharin in 1968. It stated that a portion of 1 gram or less of saccharin was not a health hazard. The report concluded that there was not enough data to make a full determination of the safety of this artificial sweetener. Further study was recommended. The NAS committee repeated its recommendations two years later.

Four years after the original NAS recommendations, the FDA decided to remove saccharin from the GRAS list. New studies were going on at the time of this decision. The FDA also established temporary limits on the use of saccharin.

Later, two new studies caused more concern about the safety of saccharin. One study was conducted by the Wisconsin Alumni Research Foundation, the other by the FDA. In the tests male and female rats were fed doses of saccharin from just after they were weaned until maturity. Both tests showed that the rats developed bladder tumors. But the manufacturers of saccharin did not accept these

*Small amounts of the controversial
sweetener saccharin (left) achieve
the same level of sweetness as do
much larger amounts of sugar (right).*

findings. They claimed that some impurities in the saccharin compound caused the cancer, not saccharin itself.

In 1974 Canada's Health Protection Branch of the Ministry of Health and Welfare began a study of saccharin. The purpose of the study was to clear up any scientific doubt as to the safety of the sweetener. The Canadian research focused on whether ortho-toluene-sulonimide (OTS), an impurity in the saccharin compound, was the cause of bladder cancer in laboratory rats.

The results of the Canadian tests were not what the manufacturers wanted. OTS was not to blame for the bladder cancer. The findings pointed to saccharin as the cause. Despite public demand for the sweetener and protests from the manufacturers, the Canadian government restricted the use of saccharin. It could be used only as a "table" sweetener, that is, added to coffee or tea or sprinkled on cereal. The reasoning was that only small amounts of the artificial sweetener would be put into these beverages, whereas larger amounts would be in soft drinks and processed food.

The FDA, using the Canadian findings as confirmation of the two American studies, also imposed restrictions on the use of saccharin. The agency revoked its 1972 approval for the use of saccharin in foods and beverages. However, the agency would allow saccharin to be sold over the counter as a table sweetener.

These measures on the part of the FDA did not settle the saccharin controversy. Public reaction was mixed. People with diabetes and those watching their weight complained about the restrictions. So did the manufacturers. Fuel was added to the saccharin fire when it was learned that rats in the Canadian tests had been fed huge quantities of diet soda containing saccharin. This revelation brought more complaints from manufacturers, who declared that the tests were invalid. Overdoses of any substance will cause health effects, they argued. Researchers explained that high dosages of suspected carcinogens are used in animal studies to produce readily detectable rates of cancer. But this explanation failed to satisfy critics. Once again, the methods and validity of testing food for safety were questioned.

Saccharin, like the cyclamates, remained a controversial sweetener. In 1977 Congress passed the Saccharin Study and Labeling Act. This law imposed a two-year moratorium on any ban on saccharin. Although the law allowed saccharin to be sold as a table sweetener, it stipulated that warning labels must be used. Such labels are still required. They advise consumers that saccharin caused cancer in animals. The law also authorized the FDA to conduct further studies of carcinogens and other toxic substances in foods, including saccharin.

Here is a warning on a packet of the artificial sweetener known as Sweet 'n' Low™, packaged by the Cumberland Packing Corporation, Brooklyn, New York: "Use of this product may be hazardous to your health. This product contains saccharin, which has been determined to cause cancer in laboratory animals."

NAS Reports on Saccharin

Pursuant to its new directive, the FDA arranged for the NAS to perform the required tests on saccharin.

The NAS subsequently issued a report in 1978 which concluded that saccharin was a carcinogen for animals but with a low potency level. Saccharin was also a potential carcinogen for human beings. The report added that impurities in the saccharin compound were not the cause of bladder tumors in animals.

A second report by the NAS, issued a year later, called for the complete overhaul of the nation's food protection laws. The NAS recommended that the FDA be given more latitude in dealing with and regulating hazardous substances. Saccharin was mentioned as a substance that needed further attention by the FDA. Saccharin is still restricted to use as a table sweetener.

Aspartame

This artificial sweetener (sold under the brand name NutraSweet™) is also the result of an accidental discovery. It was found during research on ulcer drugs. The manufacturer of this artificial sweetener, G. D. Searle and Company, along with the FDA, maintains that aspartame is one of the most thoroughly tested substances in the American food supply.

The FDA approved the use of aspartame for use as a table sweetener in 1974. But challenges from consumer groups prevented it from being sold to the public. To resolve the aspartame safety question, the FDA appointed a panel of impartial pathologists to study and, if possible, validate the Searle research reports.

In its report to the FDA, the panel concluded that aspartame did not pose a risk of brain damage that could lead to mental retardation or endocrine dysfunction. This risk had been the basis of the challenge to aspartame. But the panel recommended that approval of aspartame be withheld until more tests were conducted on the sweetener. What was needed were long-term animal tests to rule out the possibility that aspartame might cause brain tumors.

The FDA Center for Food Safety and Applied Nutrition and the FDA commissioner disagreed with the panel's recommendations. Consequently, the FDA approved the use of aspartame in carbonated beverages in 1983. Later in the same year the sweetener was approved for use as an "inactive ingredient" in human medicines.

Warning Label for Aspartame

Products containing aspartame must carry a label warning individuals who suffer from the disease called phenylketonuria (PKU), a genetic deficiency. The product label must state that high concentrations of phenylalanine, an amino acid, are present in the product. Phenylalanine is an essential human nutrient, but children with PKU cannot metabolize it, and significant intake of aspartame may result in brain damage and mental retardation.

There were more complaints about aspartame. Critics of its use charged that the sweetener broke down and exposed consumers to high levels of methanol. Excessive amounts of methanol are poisonous and can cause blindness. When methanol metabolizes into formaldehyde, it produces nasal tumors in laboratory animals. Formaldehyde is a known potent carcinogen.

Aspartame can decompose if a beverage is stored for a long time at high temperatures, according to opponents. When decomposition occurs, consumers could be exposed to methanol and formaldehyde. The FDA reviewed these and other charges and issued a statement to the effect that "the levels of methanol from aspartame in carbonated beverages did not pose any safety issues." The methanol levels, according to the FDA, were well below those that would cause health problems. The FDA report also stressed the fact that methanol was naturally present in other foods, such as juices, fruits, and vegetables. Moreover, these sources exposed consumers to higher amounts of methanol than did aspartame.

Some Reported Effects of Aspartame

A number of consumers complained of headaches, dizziness, and other symptoms after drinking beverages containing aspartame. Accordingly, the FDA asked the Centers for Disease Control in Atlanta, Georgia, to evaluate these complaints. The CDC reported that some individuals may have a high degree of sensitivity to products containing aspartame. But the data did not yield enough evidence to show that aspartame posed any serious risks to the general public. Even though a variety of symptoms were presented by the complainants, the CDC regarded them as mild. The CDC report noted that few of the complainants went to a physician because of their symptoms. The report concluded that a segment of the public might be sensitive to aspartame, and it recommended further clinical studies.

Now that cyclamates have been given a second chance, consumers have three artificial sweeteners. Questions remain about their safety. The risk-benefit factor comes into play again. And the choice is up to the consumer.

EIGHT

OTHER SOURCES OF FOOD CONTAMINATION

Food may become contaminated by broken glass, hair, animal feces, and other foreign matter. Such contamination is always of great concern to consumers and health authorities. Of equal concern is the contamination of food by microbes. Food contaminated by bacteria can cause serious illness and even death.

Bacteria That Cause Food Poisoning

Microbes most commonly associated with food poisoning are *Salmonella* species, *Staphylococcus aureus, Bacillus cereus, Clostridium perfrigens,* and *Campylobacter* species. *Salmonella* is responsible for a large percentage of food poisoning cases.

A recent study conducted by the Centers for Disease Control (CDC) revealed that in a ten-year period, 1976–1986, *Salmonella* infections increased by more than six times the rate recorded for previous decades. Scientists at the CDC have connected nearly nine thousand deaths a year to food poisoning. And Canadian estimates of food poisoning exceed five hundred thousand cases each year.

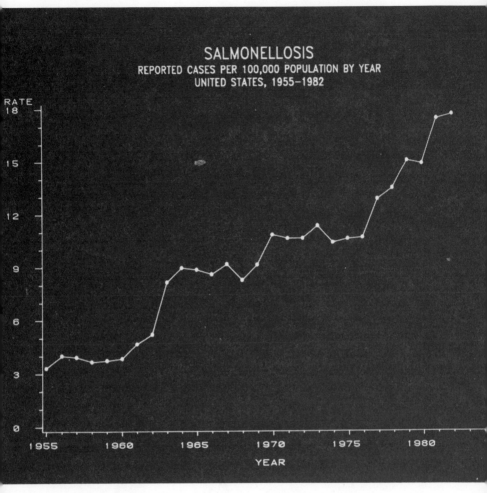

A graph illustrating the rise in the disease salmonellosis during the last thirty years

Food contaminated with *Salmonella* or other organisms can cause nausea, vomiting, diarrhea, and cramps. *Salmonella* can cause fatalities, too. Mortality is highest in infants, young children, and elderly persons, especially the sick and weak.

Salmonellosis is usually seen in mass outbreaks in schools, hospitals, and nursing homes, or at banquets and other large gatherings where people may be served contaminated food. But it can occur in the home. In general, bacterial infection of food is traceable to unsanitary handling of food and improper refrigeration. Any relaxation in sanitation standards or breakdowns in refrigerators or freezers can result in bacterial contamination of food. This is true for food traveling through the supply chain, starting with the producer and ending with the consumer.

Certain foods provide an excellent medium for the growth of food poisoning organisms. Many cases of food poisoning have been linked to contaminated eggs, chickens, prestuffed turkeys, potato salad, seafood, cream pies, various frozen foods, certain chocolates, cheese, and dried and fresh milk. In a Canadian dairy, for example, milk contaminated with *Salmonella* was used to manufacture cheese. The contaminated cheese was distributed throughout Canada, and hundreds of people became ill with salmonellosis.

Contaminated Chickens

A few years ago, millions of people watched television programs that exposed unsanitary conditions in poultry processing plants. What they saw and heard alarmed and disgusted them. Chickens were dropped into what appeared to be dirty water. Some fell onto filthy and bloody floors; these chickens were picked up without being cleaned and sent back along the processing line.

The organisms that cause food poisoning occur naturally and are hard to control. Chickens carrying bacteria arrive at the processing plant. They are brought there in cramped, dirty pens. After slaughter, the chickens are dipped into a tank of scalding water. Feces get into the tank and create a bacterial soup composed of blood, feces, and pathological organisms.

Spread of Bacteria

After their scalding bath, the chickens are plucked by a machine with mechanical fingers. This action sometimes forces any remaining feces from the carcass. Feces get pushed into the bird's skin by the force of the mechanical fingers. Once on the skin, bacteria hide in the feather follicles or spaces left by plucked feathers.

Next, the chickens are disemboweled, a process that may further spread bacteria. The eviscerating, or disemboweling, machine can rip the intestines, spilling feces and any bacteria present. The feces and bacteria may spread inside the body cavity.

True, the carcasses are washed. But the bacteria cling to the skin even when the chicken is dipped into the "chiller tank." The purpose of this dipping is to slow down the rate of spoilage. Critics of the chiller tank procedure say that dipping does something else besides slowing down spoilage: it spreads contamination. The tank eventually becomes another bowl of fecal soup after many chickens are dipped into it.

In previous years United States Department of Agriculture (USDA) inspectors would have condemned any chicken carcass that contained feces. Now the USDA allows poultry processors to wash off any feces. How thoroughly the chickens are washed is a matter of debate. It has been estimated that one of every three chickens passed by USDA inspectors is contaminated with *Salmonella* or other organisms. A National Academy of Sciences report has linked contaminated poultry to a large number of the 4 million cases of *Salmonella* and *Campylobacter* food poisoning infections that occur in the United States each year.

The Leahy Poultry Bill

A few years ago Senator Patrick Leahy of Vermont introduced a bill that called for the USDA to develop certain standards and tests for meat, poultry, fish, and animal feeds. The standards that applied to poultry plants would be enforced by the USDA.

An important provision of the Leahy bill is that a citizen would be able to sue for enforcement of the poultry inspection law if the USDA failed to act. Those persons who reported sanitation violations or who brought suit—"whistle blowers"—would be protected from harassment. There has been no action on the bill.

Residue Violation Information System

The FSIS and the FDA keep close watch on poultry as far as additives and chemical residues are concerned. The FSIS maintains what is called the Residue Violation Information System. It is a nationwide interagency computer information system. The system operates 24 hours a day, 365 days a year, and provides data on residue violations in livestock and poultry. The system is located on a USDA computer in the National Computing Center in Fort Collins, Colorado. Information is usually entered into the computer within a week after discovery. The Residue Violation Information System is a valuable tool for the discovery of hormones and antibiotics.

What Can Be Done about Contaminated Chickens?

Despite federal and state inspections of poultry, contaminated chickens go into millions of homes in the United States and Canada every year. People of all ages are exposed to *Salmonella* and other harmful organisms carried on the chicken skin and in the carcass. What, if anything, can be done to eliminate or at least reduce bacteria in our poultry supply? What are processing plants doing about the problem? And supermarkets?

Here is what Frank Perdue, president of Perdue Farms, Inc., said: "When poultry is mishandled or standard sanitation practices are ignored, bacteria can reproduce enough to affect consumers." According to Mr. Perdue, his processing plants "produce one of the cleanest birds in the industry," thanks to a vigorous quality-control program and standards. "In fact," added Mr. Perdue, "we employ five times as many quality inspectors per plant as the average producer."

Managers of meat and poultry departments in supermarkets routinely reject any obviously spoiled chickens. *Salmonella* organisms cannot be seen with the naked eye. All that can be done to prevent spoilage is proper chilling and freezing of chickens and —most important—proper cooking. Admittedly, these procedures do not eliminate bacteria from chickens. Both Frank Perdue and supermarket managers offer guidelines for the home handling of chickens.

Perdue Farms offers this advice: "Basic home sanitation practices include thorough washing of poultry, including the body cavity, before cooking. Hands, pots, pans, utensils and all counter tops and tables should be washed with soapy water after handling chicken. Finally, poultry should be thoroughly cooked."

Contaminated Eggs

Eggs, of course, are chicken products and can become contaminated with *Salmonella* organisms. Eggs are high in protein. Egg protein is so close to the perfect form of protein that nutritionists use it as a standard to measure the value of protein in other foods. Eggs are also the source of vitamin A, riboflavin, natural vitamin D, and iron. They are also high in cholesterol.

The route by which eggs become infected with *Salmonella* bacteria is still under investigation. Certainly, the crowded conditions in poultry production plants—the result of the modern factory farming system—help to foster the growth of harmful bacteria. One route of entry to eggs is by chickens consuming contaminated feed. Another may be egg contact with contaminated feces. At any rate, the *Salmonella* bacteria are transmitted to human beings by way of shell eggs and egg products.

So far there is no sure method of controlling egg

*The form egg processors use
for registering with the USDA
is shown here above a carton of eggs.*

infection on the poultry farms or plants. Therefore, efforts are directed at reducing human exposure to the harmful bacteria. Raw and undercooked eggs are the greatest sources of danger. This is especially true for sick or elderly persons, who are in the high-risk group. Raw eggs are no longer recommended as a health food, as once they were. A raw egg can be loaded with *Salmonella* or *Campylobacter* bacteria.

Because bacteria grow rapidly at room temperature, eggs should be treated as though they are always contaminated. They should be refrigerated promptly when brought home. Storage should be at a temperature below 45°F to prevent multiplication of bacteria. Foods containing eggs should be thoroughly cooked to an internal temperature of 165°F. Fried eggs can be rendered safe by frying first one side, then the other. Eggs with cracked shells should be avoided. All pans, blenders, bowls, and other utensils should be cleaned and sanitized after contact with raw eggs to prevent cross-contamination of other foods.

Seafood

Americans and Canadians consume large quantities of seafood every year. There are more than two hundred species of fish and shellfish sold in both countries. Seafood, especially fish, is recommended by nutritionists as a low-fat, nutrient-rich substitute for meat. But all is not well with the supply of food from the sea. More and more people are becoming concerned about the continued pollution of the coastal waters and bays, the sea regions that are rich in fish and shellfish. The problem of ocean pollution and the increased incidence of spoiled seafood brings up a question. Does the hazard of eating

*The continued pollution of rivers
results in bacteria that multiplies
rapidly and causes fish to die.*

polluted fish and shellfish outweigh the nutritional advantages?

All kinds of fish are sold in fresh, frozen, canned, smoked, or pickled form. Fresh fish, of course, has to be kept refrigerated to prevent spoilage. According to federal regulations, frozen fish must be solidly frozen and as hard as a brick.

Canned fish are packed in a variety of convenience and specialty items. For example, canned fish includes salmon, sardines, and tuna. Salmon species include the red (or sockeye), chinook, medium red, silver (or coho), pink, chum, and keta. Six species of tuna—albacore, blackfin, bluefin, skipjack, yellowfin, and little tuna—are packed and canned in oil or water. Cured fish includes pickled or spiced herring, salt cod and salmon, and smoked chub, salmon, and whitefish.

There are two groups of shellfish: crustaceans, such as crabs, shrimp, and lobsters; and mollusks, such as clams, oysters, and scallops. Littleneck and cherrystone clams are often served raw on the half-shell. However, in these times of ocean and bay pollution, eating raw fish or shellfish can be very dangerous. Pesticides, toxic chemicals, lead, and sewage sludge have fouled the sea environment and contaminated fish and shellfish.

Polluted Seafood

Ocean pollution is global and presents a serious threat to our seafood supply. Two groups of industrial chemicals, known as chlorinated hydrocarbons, are responsible for the widespread pollution of the sea and its resources. They are dioxins and polychlorinated biphenyls (PCBs), both very toxic chemicals.

PCBs in seafood may pose a special risk for nursing mothers. This danger was determined by a joint panel of American and Canadian scientists. The panel concluded, after a study on animals, that minimum levels of PCBs in human milk would not pose a hazard to infants. However, high levels of the chemical compound were another matter. Such levels would be dangerous. The panel's report also stated that certain women may be overexposed to PCBs, either by a high consumption of polluted seafood or by exposure at their workplace.

PCBs are produced by chlorinating biphenyl. They are used for various purposes, for example, in paints and electrical systems. They get into the environment by leakage or direct dumping. Researchers have learned that PCBs accumulate in animal tissues with pathogenic and teratogenic effects. They were first identified as pollutants in the late 1960s, at which time it was discovered that they had entered the food chain by way of contaminated seafood.

Canada has placed restrictions on the use of PCBs, with the result that the levels of the chemical compound in seafood consumed in that country have declined. In the United States, PCBs are no longer produced, and the FDA banned their use in machinery used to process food and animal feed. Because PCBs persist in soil and sediment in water, they are still a major pollutant of seafood. In the spring of 1989 high levels of PCBs were found in bluefish caught off the New Jersey coast. Bluefish are a delicacy enjoyed by many people.

Dioxins

The presence of dioxins in bays and coastal waters also contributes to the concern about the safety of our seafood. Dioxins are any of several hydrocarbon compounds that form as impurities in petroleum-derived herbicides and during some industrial processes. They are known carcinogens and teratogens. Some of the dioxin residues reach the sea by way of water runoff from farm fields. Another source is the burning of material containing this chemical. It is discharged into the air and eventually settles in the ocean and bays.

United States and Canadian health agencies have set limits on the amount of dioxins that may be present in seafood. American and Canadian scientists share information about the prevalence of both dioxins and PCBs in seafood. The polluting of the ocean environment and contamination of its food resources are major problems. The United States imports and exports fish. Thus, we may be sending contaminated fish to other countries and, because ocean pollution is global, importing seafood that is contaminated.

The National Seafood Inspection Program

Seafood inspection on the federal level is performed by the National Marine Fisheries Service (NMFS), an agency in the National Oceanic and Atmospheric Administration, United States Department of Commerce. It is designed to help the fishery industry in the harvesting, processing, and distribution of seafood.

The inspection program, however, is a voluntary one. It is provided on a fee-for-service basis. Anyone in the fishery industry, including restaurants, may request the service. A wide range of services is available. For example, NMFS inspectors will examine boats and fishing gear, processing plants, sanitation programs, storage of seafood, and maintenance of equipment.

NMFS inspectors will also determine whether a fishery product is safe, clean, and wholesome during processing and while in storage. They also inspect the labels on fish products. Like other foods, fishery products must carry proper labels. Inspectors make certain that the labels on canned, frozen, or other prepared seafood meet the requirements of the FDA.

The NMFS also operates laboratory facilities, where tests are performed to determine bacterial levels, chemical contamination, rate of decomposition, and species identification of fish.

The function, then, of the NMFS is to help the fishery industry provide a safe and wholesome supply of seafood. In cooperation with the FDA, the NMFS routinely publishes a list of seafood processing plants and their products that have been approved. When a product has been approved by the NMFS, it is so identified on the label by the NMFS logo and the notation "Packed Under Federal Inspection."

Inspection of Imported Seafood

The United States imports seafood from Japan, Norway, and other countries. The NMFS inspection program is extended to seafood importers. Their products are inspected after entry into the United States. Some of the things inspectors look for are compliance with FDA quality and safety requirements. NMFS inspectors also do pretesting and analysis of imported seafood for microorganisms,

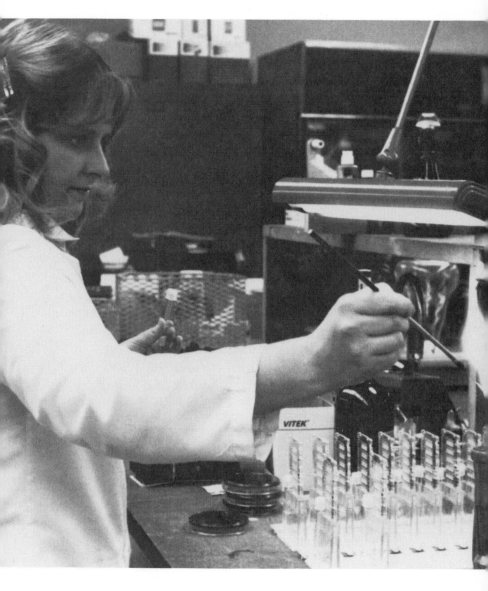

A microbiologist for the FDA prepares a test for the presence of bacteria in imported frozen shrimp.

chemical contamination, and other biological information. A check is made to see that species identification is accurate. In other words, imported tuna must be tuna, and salmon must be salmon.

State Fishery Inspections

Because the federal fishery inspection program is a voluntary one, not all people in the fishery industry take advantage of it. Therefore, most states conduct inspections along the lines of the NMFS. New Jersey, for example, is a coastal water state with a large fishery industry. The coastal waters and bays bordering this state are polluted, as are its rivers. Thus, a constant monitoring of the waters from which fish and shellfish are taken is an important function of the New Jersey Department of Environmental Protection. And the New Jersey Department of Health keeps an eye on the safety of seafood. This agency sends a bulletin to all local health departments, advising them of any source of polluted seafood.

How Effective Are
Seafood Inspections?

A special report on the risks of seafood was aired on New York's television Channel 4 in February 1989. An alarming fact brought out in the program was that nearly "75 percent of all seafood consumed in the United States goes to market without being inspected." A headline in a newspaper ad for the program warned consumers that "There's no telling what you'll catch from the catch of the day."

Critics of the seafood inspection situation claim

that some of the plants inspected by the NMFS are below the agency's standards. The FDA does not adequately police the fishery industry because of personnel shortages, pressures from commercial lobbies, and other problems, according to critics. They further charge that manufacturers' greed and carelessness in processing seafood are responsible for inferior and possibly dangerous fishery products sold to consumers.

What Are the Risks
Posed by Seafood?

In addition to the harmful effects of dioxins, PCBs, and other toxic chemicals, there is the danger of food poisoning from fish and shellfish contaminated by microorganisms. Spoiled or rotten fish or shellfish are not difficult to detect because they give off a strong, foul odor. But it is not easy to identify spoiled seafood once it has been broiled, baked, fried, or pickled. The same is true if it is canned or smoked. At least, the average consumer would have difficulty in detecting spoilage.

Restaurant and
"Takeout" Seafood

Although most restaurants and other retail sellers of seafood maintain sanitation standards, there are always breakdowns in freezers and refrigerators. Employees become careless. As a consequence of equipment breakdown and relaxation of sanitation standards, customers may be served spoiled fish. Unfortunately, the customer has no control in such a situation. If the seafood is obviously spoiled, judging by the odor and taste, a customer can reject it. When it is not so obvious, he or she is at risk.

One popular seafood restaurant chain, Red Lobster, states on its table mats that its employees—from cook to server—"are trained in the principles of quality control." All seafood entering the Red Lobster is, according to the mat information, inspected by trained managers to ensure that it is safe and wholesome.

Surimi

A favorite seafood in Japan for more than nine centuries is surimi. It is now a popular seafood item in America. Surimi is a compound made from fish flesh, usually pollack from Alaskan waters. The pollack flesh is minced and washed in cool water until it turns into a thick paste. The fish paste is then mixed with various flavorings, preservatives, and stabilizers, including sugar, salt, and cornstarch. Monosodium glutamate is often added to the mix. It is then shaped into blocks and quickly frozen.

Surimi is often offered as crab, shrimp, lobster, fish, or scallops at some salad bars and stores. This type of sale is deceptive and illegal. Surimi is made to resemble crab and lobster by artfully molding the mixture into strips or chunks. The strips or chunks are then colored with streaks of red food coloring to resemble crab or lobster meat. Seafood gourmets and fish experts can tell the difference, though. Surimi has a rubbery texture and a salty-sweet taste.

Pure-food advocates charge that the public is being duped by the surimi processors and distributors. A study conducted by the National Food Processors Association found that surimi has a low fat and cholesterol content, but it also has little protein and a high sugar and salt content.

A surimi mixture that weighs 4 ounces contains about 735 milligrams of sodium. This amount of salt is much higher than that of real scallops. And according to critics, the water-washing process rinses out a considerable amount of vitamins and minerals. It is also claimed that the manufacturing process eliminates a valuable fish oil known as Omega 3, a fatty acid that is believed to protect people from heart disease.

Surimi processors admit that their product has some nutritional deficiencies. But, they say, surimi, a processed seafood product, is comparable to processed cheese. It's not the real thing, but it tastes like it. And they point out that surimi has more protein than yogurt or processed meats.

The FDA, in a 1985 ruling, specified that packaged surimi must be labeled as "imitation" unless it has been fortified with nutritional additives equivalent to natural crab, scallops, or other seafood. But these restrictions are sometimes ignored by food stores and fish markets.

The Nature of Seafood

Seafood is a highly nutritious but very perishable food source. Because it is so perishable and subject to contamination by bacteria and chemicals, the FDA and NMFS have enacted some strict rules and regulations. They are aimed at safeguarding the wholesomeness of our food supply from the sea. How well these two agencies protect the public from contaminated or unwholesome seafood is debatable, according to critics. Despite the criticism, the NMFS does not believe that our seafood is as bad or as hazardous as some consumer groups charge.

True, there are a number of cases of illness each year caused by contaminated seafood. But the NMFS believes that 90 percent of these cases can be traced to toxins in some species of tropical fish and to mollusks illegally harvested from waters known to be polluted. Another source of illness results from improper handling of fish and shellfish in stores, restaurants, and home kitchens.

The American fishing industry—already faced with stiff harvesting competition from Japanese, Russian, and other foreign groups—wants no more problems. Its members are concerned about the pollution of our seafood, but they do not think the situation is as bad as people are being led to believe. They also think the public is needlessly alarmed. Thus, just as in the case of other problems in the food chain, consumers must decide who is right about the safety of our seafood. And consumers must decide whether the benefits of eating seafood outweigh the health risks.

NINE

FOOD IRRADIATION

The FDA has approved the limited irradiation of food, a cause for alarm among pure-food advocates. By 1980 the FDA had permitted the irradiation of wheat, wheat flour, potatoes, and about sixty herbs and spices. Six years later, pork, fruit, and vegetables were added to the list. However, there was strong opposition to the irradiation of any food. Consequently, only herbs and spices are being irradiated at present. But pressures are building to begin irradiation of a wide range of foods, including troublesome poultry.

What Is Food Irradiation?

In irradiation, food is exposed to radioactive materials, cobalt 60 and cesium 137, or to X-ray generators. Cobalt 60 yields 2 gamma rays of 1.17 and 1.33 million volts (MeV). Cesium 137 produces 0.66 MeV of gamma radiation. These levels are below the recommended maximum of 10 MeV and are therefore considered safe.

Irradiation of food rearranges the molecular structure of macromolecules in bacteria and fungi. The process splits the chemical bonds in the molecules with high-energy rays to form ions and free radicals. When enough of the critical bonds have been split in harmful organisms in food, the organisms are killed. However, food is also altered during the irradiation process. And as we shall see, that is one of the numerous objections to irradiation of food.

Irradiation plants have specially built rooms with walls 6 to 8 feet thick. A conveyor moves food past the irradiation beam. The dosage is determined by how long the food is exposed to the rays. Dosage is stated in rads or kilograys. Depending on the food, irradiation doses range widely.

Different foods require different doses. Wheat and wheat flour require a dosage of 50,000 rads, or 0.5 kGy. Potatoes are given a dose of 15,000 rads, or 0.1 kGy (to prevent sprouting). Pork can be treated with irradiation to control wormlike parasites called trichinae. The recommended dosage is 100,000 rads, or 1 kGy. Fruits and vegetables are treated for insect control and receive up to 100,000 rads, or 1 kGy. Herbs, spices, and vegetable seasonings can be irradiated with dosages of 3,000,000 rads, or 20 kGy.

Although only herbs and spices are being irradiated in the United States at this writing, a number of foreign countries irradiate a variety of foods, including potatoes, onions, strawberries, mushrooms, bananas, various vegetables, fish, pork, and pork products.

In Canada, food irradiation is regulated by the food additive section of the food and drug regulations established by the Health Protection Branch of the Ministry of Health and Welfare. Although irradiation is classified as an additive, there is pressure in Canada and the United States to reclassify irradiation as a process. Such a classification would eliminate preclearance requirements and other restrictions now placed on additives.

Why Food Irradiation?

First of all, food irradiation is not a new process or treatment. Swedish food processors used the method on strawberries in 1916, and patents for food irradiation were taken out in the United States in 1920.

Supporters of food irradiation state that the process can be an important technological advance in the sterilization and preservation of the food supply. It is supposed to have the following advantages:

It kills pathogenic organisms.
It extends the shelf life of food products.
It prevents the sprouting of potatoes and vegetables.
It retards the ripening of fruits, thus aiding in shipping and storage.
It controls insects in stored food and therefore can reduce or eliminate the need for postharvest pesticides.
It can be used to tenderize meat and improve the quality of some foods.

However, some critics have suggested that spoiled food could be sterilized by irradiation and then offered for human consumption. Such a use would be a cause for alarm. The United Nations World Health Organization (WHO) prohibits such a deceptive use of food irradiation.

Additive or Process?

If and when food irradiation is approved by the FDA for wider use (and the indications are that it may be), the question of whether it is a food additive or process will have to be resolved. Under present FDA regulations, food irradiation is considered to be an additive. As such it is subject to the same restrictions and regulations as the chemical additives. But irradiation supporters argue that it is a process and should be exempt from additive restrictions. They complain that the Delaney amendment to the Food, Drug and Cosmetic Act, which included food irradiation in the definition of an additive, is too restrictive.

Food irradiation advocates also argue that the classification of food irradiation as an additive is an unwarranted barrier. It prevents the wider use and acceptance of this beneficial preservation technique. If it were reclassified, the extent of testing required to prove that irradiated foods are safe would be reduced. The FDA position, however, is that, whether process or additive, food irradiation would require the same extensive animal studies before being approved.

Long-term feeding studies of irradiated food were conducted by the United States Army. An outside laboratory was hired to do the testing. Laboratory animals were fed pork, ham, and beef. The FDA informed the army that the Delaney amendment did not require that every irradiated food be individually tested for safety. What this meant was that if tests on irradiated beef, pork, and chicken showed that they were safe, then all products irradiated in that food group would be considered safe. Unfortu-

nately, the methods used by the contracted laboratory were not adequate. And after twenty-five years
of research on food irradiation, at an approximate
cost of $51 million, the army has not used irradia-
tion in the military food system.

WHO established a committee to investigate the
safety and wholesomeness of irradiated foods. In its
report the committee defined irradiation as a physi-
cal process for treating food. As such, the irradiation process is comparable to heating or freezing
food. The committee recognized that the use of
radioactive materials to treat food is controversial.
It is a matter that has drawn the attention of consumer groups and many concerned people. Nevertheless, WHO has concluded that food irradiation is
safe when the proper dosages are used. The European Economic Community's Scientific Committee
agrees with this assessment. The FDA is not sure
about it.

Is Food Irradiation Safe?

It depends on who answers the question. Some scientists say it is safe. Others have reservations. And
consumer protection organizations say food irradiation is hazardous and unnecessary. The National
Coalition to Stop Food Irradiation, a San Francisco–
based organization, states in its literature that "we
believe that irradiated food is unhealthy and unsafe
to consume." Other American and Canadian consumer protection organizations take the same position.

According to the National Coalition to Stop Food
Irradiation, laboratory animals fed irradiated foods
developed testicular tumors, kidney ailments, short-

ened life spans, weight loss, and an increased rate of infertility and death of their offspring.

Opponents of food irradiation believe that widespread treatment of food with radioactive materials will open a Pandora's box of evils. The curious Pandora of mythology opened a forbidden box and released all kinds of disease and evils into the world. One evil of food irradiation, say those who are against the process, is that it may deplete or destroy the vitamin content of food. They also state that there is a breakdown of amino acids—tryptophan, cysteine, and phenylalanine—all of which are important in human nutrition. Furthermore, irradiation may cause carbohydrates to form toxic chemicals, according to opponents of the process. And nucleic acids and enzymes may also be adversely affected by the process.

Safe-food advocates point out that aflatoxin, a naturally formed food toxin produced by molds, is produced in greater quantities in some irradiated foods. Some aflatoxins have caused cancer in test animals. Another drawback is that botulism, a deadly form of food poisoning caused by the growth of *Clostridium botulinum* in canned foods, would still be a threat to consumers. Irradiation dosages now approved by the FDA would not kill the organisms.

There is also the possibility that irradiation could cause mutations of disease-bearing organisms. Scientists are concerned about a new group of chemical compounds that can be produced in irradiated foods. These chemicals are unusual substances known as radiolytic products (RPs). They are not found in untreated foods. RPs may be carcinogens or mutagens. Many of them are unique chemical compounds about which little is known.

The FDA admits that food irradiated with minimum dosages could contain RPs. More studies are needed, especially if higher dosages are to be used. Foods irradiated above 100 rads, or 1 kGy, should be tested for toxicity. This means more animal studies.

Radioactive Food?

Another concern voiced by pure-food advocates is that irradiated food might become radioactive itself. They state that when the energy level rises above 10 to 15 MeV, significant amounts of radioactivity can result. Thus, it is important to keep radiation levels low in the treatment of food. Consumer groups question whether this can be done all the time. Also, they say that any damage to the source of the irradiation can contaminate food to a dangerous degree.

Food irradiation supporters do not agree. They argue that food will not become radioactive unless it contains traces of silver, tin, barium, or strontium. Or unless there is human error or equipment failure. And these are factors that must be considered. Finally, supporters claim that if irradiated food is stored for a time before being consumed, the radiation level will be low or even insignificant.

Possible Environmental Problems

If a large number of food irradiation plants were allowed to operate in the United States—and it would indeed take thousands of them to handle the huge American food supply—there is the possibility of accidents. Radioactive materials could be discharged into the environment. The nuclear accidents at Three Mile Island in Pennsylvania and

EPA technicians taking a deep-soil sample to determine possible dioxin contamination

Chernobyl in the Soviet Union are reminders of what could happen. Irradiation proponents advise that electronic beam irradiation, instead of radioactive material, could be used.

There is another factor to consider. Cesium 137 is stored in a water-soluble form. Any leak into the groundwater supply could irreversibly contaminate our water resources and work up into the human food chain. Another possible hazard would result from trucking accidents. Trucks hauling radioactive materials to irradiation plants could have accidents that could release radioactive materials onto the highways and into the environment. Food irradiation promoters say that such considerations are overemphasized by opponents of irradiation.

Who Promotes Food Irradiation?

In the United States the chief promoters of food irradiation are the food industry, the United States Department of Energy (DOE), and private contractors who would gain from the widespread use of food irradiation. Food irradiation research is being funded by the DOE, an agency that wants to find some use for its surplus nuclear materials. The DOE has plans to build food irradiation plants in Hawaii, Alaska, Washington, Iowa, Oklahoma, and Florida.

Food irradiation is also a controversial subject in Canada. There is no irradiated food being sold in Canada at present, unless it is on the black market. However, Canada is a world leader in food irradiation technology. The technology is regulated by the Crown Corporation Atomic Energy of Canada Ltd. The technology is made available to other countries. Canada also provides more than 90 percent of the world supply of cobalt 60 for food and medical irradiation plants. There are about fifty such plants around the world.

Canadian opponents of food irradiation, like those in the United States, maintain that irradiation is not necessary. For one thing, it is too expensive when compared to other ways of preserving food. For another, the technology cannot be controlled because there are no adequate tests to determine whether a food has been irradiated and at what dosage. And the Canadian opponents say that Canada should not be promoting food irradiation abroad, as it has been doing in Thailand and Jamaica.

The International Atomic Energy Agency (IAEA) would like to see developing countries erect food irradiation plants. By way of encouraging such projects, the IAEA suggested that developing countries would find markets for their irradiated food products in Europe and North America. In 1988 the IAEA sponsored a conference in Geneva, Switzerland. The agenda included consideration of ways to increase consumer acceptance of food irradiation.

Food irradiation promoters keep stressing the advantages of this controversial process. They believe that any process that will eliminate or reduce the use of harmful chemicals is a good thing for consumers. They stress the fact that food poisoning is increasing despite modern food storage and handling techniques. In fact, they say that in some parts of the country, people eat seafood at their peril.

On the other hand, food irradiation opponents disagree with these statements. Unsafe seafood does not result just from spoilage; it is caused also by contamination from polluted coastal waters, bays, rivers, and lakes. Irradiated food will not chase away the gnawing hunger of starving people around the world. Hunger is not the result of food processing or preservation techniques. It stems from economic and political considerations.

Also, opponents doubt whether widespread food irradiation will reduce dependence on pesticides and other agricultural chemicals. They point to the fact that food is irradiated after harvest, not before. Farmers must still grow the food to be irradiated. As for irradiation's reducing the need for chemical additives, opponents say that is questionable, too. They believe that some food processors would add chemicals to offset changes in food texture, flavor, and odor caused by irradiation.

Inspection of Irradiated Food

At present, there is no test that will detect irradiation of food. Nor is there any way of verifying whether recommended dosages of radiation have been used. Consumers would have to take their irradiated food on trust. Many people will no longer accept their food on trust. The laxity of the EPA and FDA in taking action on behalf of consumers, as well as the questionable ethics of some manufacturers, are among the reasons many people doubt the safety of their food supply. And because of the growing public mistrust and alarm about food irradiation, supermarkets say they will not stock known irradiated foods, at least not until consumer fears have been eased.

Labeling of Irradiated Food

If and when food irradiation becomes more widespread in the United States, the only labeling re-

quired will be a reproduction of the irradiation symbol. The symbol is a flower, known by the coined name of "radura." The radura will tell consumers that the food has been irradiated. The food industry will accept the symbol but not any labeling that spells out "Irradiated Food Product."

Food protection groups insist, however, that the food industry and federal government must allow clear and prominent labels. The radura logo may not be enough; unless people learn to recognize it and what it stands for, its intent will be lost. Consumers must have a clear choice of which food they want to purchase and consume.

Food irradiation and its labeling were contested subjects in the New Jersey legislature a few years ago. Several anti-irradiation bills were introduced into the legislature. One opposed the irradiation of food for New Jersey consumers. The other called for specific labeling of irradiated food in retail stores and restaurants. Both bills were opposed by the Food Irradiation Service of Isomedix, Inc., Whippany, New Jersey. The lobbying efforts of this corporation helped to defeat the two bills.

In November 1989 the sale of irradiated foods was banned in New York State, with the exception of spices and foods served to hospital patients with immune-system deficiencies. Maine has also banned the sale of irradiated foods.

Consumer Right to Know

Safe-food organizations believe that a consumer has the right to know if a food product or restaurant meal or takeout food has been irradiated. Irradiated foods are, in a sense, camouflaged. They look fresh and wholesome. Although they may be wholesome or edible, they may not be fresh. Because there may be deception, intentional or not, consumer advocates demand that labels and menus clearly state that the food has been irradiated.

Labeling food as being irradiated seems to be poison to some food processors, distributors, and restaurants. They claim that when wider use of the process has been approved by the FDA, glaring labels will deter consumers. And in addition to defining irradiation as a process rather than an additive, manufacturers would like another name or term for it.

Nutritionists urge Americans and Canadians to change their dietary habits for better health. People are advised to eat less fat, more roughage, and fish instead of red meats. More roughage means consuming more grains, such as oats and wheat. It means eating more leafy vegetables and fresh fruit. But these are the very foods that will be irradiated if the FDA approves wider use of the treatment.

Fish, of course, will be a prime candidate for irradiation because of its high spoilage rate. So will chicken, with its tendency to support the growth of harmful bacteria. Thus, an important question arises: will the advice to eat more healthy foods be negated by the loss of valuable nutrients and vitamins through irradiation?

Other questions about the safety of food irradiation crop up. For example, can consumers depend on manufacturers to use the proper dosage of irradiation? What is the possibility that this method of food preservation will create mutant and resistant bacteria? Will food irradiation cause changes in white blood cells? Will it cause lower birth rates? And will longtime consumption of irradiated foods cause tumors in human beings?

These questions and those prompted by the widespread use of food additives, hormones, and antibiotics, by the prevalence of pesticide residues in food, by the increase in food poisoning, and by the condoned use of artificial sweeteners will have to be answered if consumers are to have complete faith in the safety of their food supply.

Equally important in the minds of consumer advocates is public confidence in federal and state agencies responsible for the safety of our food supply. It may be that those agencies will have to be overhauled if they are to regain the public trust. The FDA, it seems, may no longer have that trust.

SOURCES

Americans for Safe Food, Washington, D.C.

Author's personal experience as a food inspector in the Veterinary Service, United States Army.

Canadian Ministry of National Health and Welfare, Health Protection Branch, Ottawa, Canada.

Coalition for Alternatives in Nutrition and Health Care, Richlandtown, Pennsylvania.

College of Medicine and Dentistry, Rutgers University, Department of Preventive Medicine, Newark, New Jersey.

Consumers United to Stop Food Irradiation, Ilderton, Ontario, Canada.

Food and Water Inc., Denville, New Jersey.

International Institute of Concern for Public Health, Toronto, Canada.

National Academy of Sciences, Washington, D.C.

National Coalition to Stop Food Irradiation, San Francisco, California.

National Fisheries Inspection Service, U.S. Department of Commerce, Washington, D.C.

Pennsylvania Department of Health, Harrisburg, Pennsylvania.

Pennsylvania State University, College of Agriculture, University Park, Pennsylvania.

Perdue Farms, Inc., Salisbury, Maryland.

United States Department of Agriculture, Food Safety and Inspection Service, Beltsville, Maryland.

United States Food and Drug Administration, Rockville, Maryland.

United States General Accounting Office, Washington, D.C.

BIBLIOGRAPHY

Books

Bernarde, Melvin A. *The Chemicals We Eat.* New York: American History Press, 1971.

Carson, Rachel. *Silent Spring.* Boston: Houghton Mifflin, 1987.

Feingold, B. F. *Why Your Child Is Hyperactive.* New York: Random House, 1985.

Jacobson, Michael F. *Eater's Digest: The Consumer's Factbook of Food Additives.* New York: Doubleday, 1972.

Schell, Orville. *Modern Meat: Antibiotics, Hormones and the Pharmaceutical Farm.* New York: Random House, 1984.

Sinclair, Upton. *The Jungle.* Urbana, Ill.: University of Illinois Press, 1988.

Turner, James S. *The Chemical Feast.* New York: Grossman, 1970.

United States Department of Agriculture. *Food for Us All: The 1982 Yearbook of Agriculture.* Washington, D.C.: U.S. Government Printing Office, 1982.

Webb, Tony, Tim Lang, and Kathleen Tucker. *Food Irradiation: Who Wants It?* Rochester, Vt.: Thorson Publishers, 1987.

Wiley, Harvey H. *An Autobiography.* Indianapolis: Bobbs Merrill, 1930.

Articles

"Advertisement: Our Food Supply Is Safe!" *New York Times,* 5 April 1989, 11(A).

"Bad Apples." *Consumer Reports,* May 1989, 288–296.

Belsie, Laurent. "Hormone Use Debate Heats Up." *Christian Science Monitor,* 17 October 1989, 7.

Bhaskaram, C., and G. Sadasivan. "Effects of Feeding Irradiated Wheat to Malnourished Children." *American Journal of Clinical Nutrition* 28 (February 1975): 130–135.

Campbell-Platt, G. "The Food We Eat." *New Scientist* 118 (May 1988): 4.

"Editorial: The Sooner, the Better." *Food and Justice Magazine of the United Farm Workers of America, AFL-CIO* 5, no. 5 (December 1988): 2.

"The Great Global Food Fright." *U.S. News and World Report,* 27 March 1989, 56–58.

"Scientists Declare Hormones Safe." *New Scientist* 115 (September 17, 1987): 29.

"Warning: Your Food, Nutritious and Delicious, May Be Hazardous to Your Health." *Newsweek,* 27 March 1989, 16–19.

Wright, R. "Ban of Cattle Hormones Cannot Be Policed." *New Scientist* 115 (September 17, 1987): 29.

INDEX

State responsibility, 79, 134
Stauber, John, 97
Supplementary nutrients, 59
Surimi, 136–137
Syntex, 94

Thalidomide tragedy, 43–44
Toxicity of pesticides, 66–67,
72–73, 78
Trenbolene acetate, 94

Uniroyal, 75
United States Army, 142–143
United States Department of
Agriculture (USDA), 90, 94,
123

Upjohn Company, 95

Verrett, Jacqueline, 108

Western Packers and Canned
Goods Association, 25
White, William Allen, 23
Wholesome Poultry Products
Act, 100–102
Wiley, Harvey H., 21, 22, 23,
25–31
Wisconsin Alumni Research
Association, 111
World Health Organization
(WHO), 51, 141, 143
World War II, 67

ABOUT THE AUTHOR

J. J. McCoy is a naturalist and noted author of seventeen books for young people in the fields of natural history, ecology, wildlife conservation, animal behavior, environmental protection, and problems of the sea. He is a graduate of Pennsylvania State University.